Discharge Measurements at Gaging Stations

By D. Phil Turnipseed and Vernon B. Sauer

Techniques and Methods 3–A8

U.S. Department of the Interior
U.S. Geological Survey

U.S. Department of the Interior
KEN SALAZAR, Secretary

U.S. Geological Survey
Marcia K. McNutt, Director

U.S. Geological Survey, Reston, Virginia: 2010

For more information on the USGS—the Federal source for science about the Earth, its natural and living resources, natural hazards, and the environment—visit *http://www.usgs.gov* or call 1–888–ASK–USGS.

For an overview of USGS information products, including maps, imagery, and publications, visit *http://www.usgs.gov/pubprod*.

To order this and other USGS information products, visit *http://store.usgs.gov*.

Suggested citation:
Turnipseed, D.P., and Sauer, V.B., 2010, Discharge measurements at gaging stations: U.S. Geological Survey Techniques and Methods book 3, chap. A8, 87 p. (Also available at *http://pubs.usgs.gov/tm/tm3-a8/*.)

Preface

This series of manuals on techniques and methods (TM) describes approved scientific and data-collection procedures and standard methods for planning and executing studies and laboratory analyses. The material is grouped under major subject headings called "books" and further subdivided into sections and chapters. Section A of book 3 is on surface-water techniques.

The unit of publication, the chapter, is limited to a narrow field of subject matter. These publications are subject to revision because of experience in use or because of advancement in knowledge, techniques, or equipment, and this format permits flexibility in revision and publication as the need arises. Chapter A8 of book 3 (TM 3–A8) deals with discharge measurements at gaging stations. The original version of this chapter was published in 1969 as U.S. Geological Survey (USGS) Techniques for Water-Resources Investigations, chapter A8 of book 3. New and improved equipment, as well as some procedural changes, have resulted in this revised second edition of "Discharge measurements at gaging stations."

This edition supersedes USGS Techniques of Water-Resources Investigations 3A–8, 1969, "Discharge measurements at gaging stations," by T.J. Buchanan and W.P. Somers, available at *http://pubs.usgs.gov/twri/twri3a8/*, and supplements USGS Water-Supply Paper 2175, volume 1, 1982, "Measurement and computation of streamflow: Measurement of stage and discharge," by S.E. Rantz and others, available at *http://pubs.usgs.gov/wsp/wsp2175/html/WSP2175_vol1.html*.

This revised second edition of "Discharge measurements at gaging stations" is published online at *http://pubs.usgs.gov/tm/tm3-a8/* and is for sale by the U.S. Geological Survey, Science Information Delivery, Box 25286, Federal Center, Denver, CO 80225.

Contents

Figures

Tables

Conversion Factors

Multiply	By	To obtain
Length		
inch (in.)	2.54	centimeter (cm)
inch (in.)	25.4	millimeter (mm)
foot (ft)	0.3048	meter (m)
mile (mi)	1.609	kilometer (km)
Area		
acre	4,047	square meter (m^2)
acre	0.4047	hectare (ha)
acre	0.4047	square hectometer (hm^2)
acre	0.004047	square kilometer (km^2)
square foot (ft^2)	929.0	square centimeter (cm^2)
square foot (ft^2)	0.09290	square meter (m^2)
Volume		
gallon (gal)	3.785	liter (L)
gallon (gal)	0.003785	cubic meter (m^3)
gallon (gal)	3.785	cubic decimeter (dm^3)
million gallons (Mgal)	3,785	cubic meter (m^3)
cubic foot (ft^3)	28.32	cubic decimeter (dm^3)
cubic foot (ft^3)	0.02832	cubic meter (m^3)
acre-foot (acre-ft)	1,233	cubic meter (m^3)
acre-foot (acre-ft)	0.001233	cubic hectometer (hm^3)
Flow rate		
acre-foot per day (acre-ft/d)	0.01427	cubic meter per second (m^3/s)
acre-foot per year (acre-ft/yr)	1,233	cubic meter per year (m^3/yr)
acre-foot per year (acre-ft/yr)	0.001233	cubic hectometer per year (hm^3/yr)
foot per second (ft/s)	0.3048	meter per second (m/s)
cubic foot per second (ft^3/s)	0.02832	cubic meter per second (m^3/s)
cubic foot per second per square mile [(ft^3/s)/mi^2]	0.01093	cubic meter per second per square kilometer [(m^3/s)/km^2]
cubic foot per day (ft^3/d)	0.02832	cubic meter per day (m^3/d)
gallon per minute (gal/min)	0.06309	liter per second (L/s)
gallon per day (gal/d)	0.003785	cubic meter per day (m^3/d)
gallon per day per square mile [(gal/d)/mi^2]	0.001461	cubic meter per day per square kilometer [(m^3/d)/km^2]
million gallons per day (Mgal/d)	0.04381	cubic meter per second (m^3/s)
million gallons per day per square mile [(Mgal/d)/mi^2]	1,461	cubic meter per day per square kilometer [(m^3/d)/km^2]
Mass		
pound, avoirdupois (lb)	0.4536	kilogram (kg)

Multiply	By	To obtain
Pressure		
atmosphere, standard (atm)	101.3	kilopascal (kPa)
pound-force per square inch (lbf/in^2)	6.895	kilopascal (kPa)
pound per square foot (lb/ft^2)	0.04788	kilopascal (kPa)
pound per square inch (lb/in^2)	6.895	kilopascal (kPa)
Density		
pound per cubic foot (lb/ft^3)	16.02	kilogram per cubic meter (kg/m^3)
pound per cubic foot (lb/ft^3)	0.01602	gram per cubic centimeter (g/cm^3)

Temperature in degrees Celsius (°C) may be converted to degrees Fahrenheit (°F) as follows:

$$°F=(1.8×°C)+32$$

Temperature in degrees Fahrenheit (°F) may be converted to degrees Celsius (°C) as follows:

$$°C=(°F-32)/1.8$$

Vertical coordinate information is referenced to the insert datum name (and abbreviation) here for instance, "North American Vertical Datum of 1988 (NAVD 88)."

Horizontal coordinate information is referenced to the insert datum name (and abbreviation) here for instance, "North American Datum of 1983 (NAD 83)."

Elevation, as used in this report, refers to distance above sea level. Because this report is based on a large number of previously published scientific investigations, "sea level" is not referenced to a single vertical datum. "Mean sea level" also is not used with reference to a single datum; where used, the phrase means the average surface of the ocean as determined by calibration of measurements at tidal stations.

Abbreviations and Acronyms

AA	Price AA current meter
ADCP	acoustic Doppler current profiler
ADC	acoustic digital current meter
ADV	acoustic Doppler velocimeters
Aquacalc	JBS Instruments Aquacalc Pro Discharge Measurement Computer
C1	C1 connector
CMD	current meter digitizer
CMCsp	Hydrological Services Current Meter Counter signal processor
CSG	crest stage gage
CST	Central Standard Time
C type	Columbus type
DCP	Data Collection Platform
DGPS	global positioning system with differential corrections
dB	decibels
EDT	Eastern Daylight Time

EFN	electronic field notebook
EST	Eastern Standard Time
FlowTracker	SonTek/YSI FlowTracker Handheld acoustic Doppler velocimeter
GPS	global positioning system
HIF	U.S. Geological Survey Hydrologic Instrumentation Facility
HWM	high water mark
Hz	hertz
ISO	International Organization for Standardization
kHz	kilohertz
LEW	left edge of water
MHz	megahertz
NAD 83	North American Datum of 1983
NAVD 88	North American Vertical Datum of 1988
NWIS	U.S. Geological Survey National Water Information System
OFR	U.S. Geological Survey Open-File Report
PDA	portable digital assistant
PFD	personal flotation device
QA/QC	quality assurance and quality control
RD	Teledyne RD Instruments
REW	right edge of water
RP	reference point
RTK	real-time kinematic
RTK-GPS	real-time kinematic global positioning system
SNR	signal-to-noise ratio
SWAMI	surface water measurement and inspection
TM	U.S. Geological Survey Techniques and Methods
TRDI	Teledyne RD Instruments
TWRI	U.S. Geological Survey Techniques of Water-Resources Investigations
USGS	U.S. Geological Survey
WAAS	Wide Area Augmentation System
WRD	Water Resources Division/Discipline
WSC	Water Science Center
WSCan	Environment Canada Ministry Water Survey of Canada Agency

Definitions of Symbols

a_i	cross-section area for the ith segment of the n segments into which the cross section is divided
arctan	arctangent
b	width of the throat section (in units of distance)
b_i	distance from initial point to location i
$b_{(i-1)}$	distance from initial point to preceding location
$b_{(i+1)}$	distance from initial point to next location
d_i	depth of water at location i
C	coefficient of discharge

c	speed of sound (in distance per unit time)
cos	cosine of the angle alpha
°C	temperature in degrees Celsius
$\mathbf{CORR}_{wl}$	correction (in units of distance) to subtract from the wet-line depth to obtain the vertical depth
$\mathbf{D}_{wl}$	wet-line depth (in units of distance)
$\mathbf{D}_{v}$	vertical depth (in units of distance)
	average change in stage in the reach L during the measurement (in units of distance)
	elapsed time during measurement
	computed phase difference
°F	temperature in degrees Farenheit
H	weighted mean gage height (in units of distance)
h	static head, head, elevation, or gage height (in units of distance)
IQR	interquartile range
L	length of reach between measuring section and control (in units of distance)
m	difference (in feet) between the air-line correction for the sounding position and that for the 0.8 position
n	difference (in feet) between the wet-line correction for the sounding position and that for the 0.8 position, if the depths are greater than 40 feet and the change in vertical angle is more than 5 percent
P	vertical angle (in degrees)
	ratio of the circumference to the diameter of a circle; approximately equal to 3.14159
Q	discharge (in volume per unit time)
$Q1$	25 percent of samples are less than this value
$Q3$	75 percent of samples are less than this value
q_i	discharge through partial section i, (in volume per unit time)
R	number of rotor revolutions per second
T	total time for the measurement
t_1	duration of time intervals between breaks in the slop of the gage height graph
tan	tangent
	time lag between pulses
V	velocity (in distance per unit time)
Vx	velocity (in distance per unit time) in the x direction (perpendicular to the tag line)
Vy	velocity (in distance per unit time) in the y direction (parallel to the tag line)
V	volume of water in container
v_i	mean velocity (in units of distance per unit time) of the flow normal to the ith segment, or vertical
W	average width of stream between measuring section and control (in units of distance)
W_1	weight of empty container (in units of mass)
W_2	weight of container with water (in units of mass)
w	unit weight of water (in units of mass per volume)
x	x-direction (perpendicular to the tag line) or x-component of velocity

Discharge Measurements at Gaging Stations

By D. Phil Turnipseed and Vernon B. Sauer

Abstract

The techniques and standards for making discharge measurements at streamflow gaging stations are described in this publication. The vertical axis rotating-element current meter, principally the Price current meter, has been traditionally used for most measurements of discharge; however, advancements in acoustic technology have led to important developments in the use of acoustic Doppler current profilers, acoustic Doppler velocimeters, and other emerging technologies for the measurement of discharge. These new instruments, based on acoustic Doppler theory, have the advantage of no moving parts, and in the case of the acoustic Doppler current profiler, quickly and easily provide three-dimensional stream-velocity profile data through much of the vertical water column. For much of the discussion of acoustic Doppler current profiler moving-boat methodology, the reader is referred to U.S. Geological Survey Techniques and Methods 3–A22 (Mueller and Wagner, 2009).

Personal digital assistants (PDAs), electronic field notebooks, and other personal computers provide fast and efficient data-collection methods that are more error-free than traditional hand methods. The use of portable weirs and flumes, floats, volumetric tanks, indirect methods, and tracers in measuring discharge are briefly described.

Purpose and Scope

The U.S. Geological Survey (USGS) makes tens of thousands of streamflow measurements each year across the United States and its territories. Measured discharges range from a trickle in a small ditch or stream [less than 0.01 cubic foot per second (ft^3/s)], to a flood on the Mississippi River (greater than 1,800,000 ft^3/s). Several methods are used by the USGS to make streamflow measurements. Principally, the USGS uses mechanical current meters and hydroacoustic meters [for example, acoustic Doppler current profilers (ADCPs) and acoustic Doppler velocimeters (ADVs)]. The purpose of this report is to describe the equipment and procedures used by the USGS and others for making discharge measurements, and to describe new developments in equipment and procedures. Other traditional methods of measuring streamflow include portable weirs and flumes, and volumetric, float, indirect, and tracer measurements. Relatively new developments include the use of a moving boat with the ADCP (Mueller and Wagner, 2009), the wading rod mounted ADV, electromagnetic current meters, electronic field notebooks, personal digital assistants (PDAs), and various procedural changes.

The original version of USGS Techniques of Water-Resources Investigations book 3, chapter A8 (TWRI 3–A8), by Buchanan and Somers (1969), was used as an extensive resource in the preparation of this publication because much of the equipment and techniques described by Buchanan and Somers are still applicable to current streamgaging methods. The USGS publications "Measurement and Computation of Streamflow, volumes 1 and 2," by Rantz and others (1982); "Discharge measurements using a broad-band acoustic Doppler current profiler," by Simpson (2002); "Quality-assurance plan for discharge measurements using acoustic Doppler current profiler," by Oberg and others (2005); and "Measuring discharge with acoustic Doppler current profilers from a moving boat," by Mueller and Wagner (2009), were also used extensively in the preparation of this publication. Numerous parts of this chapter were taken verbatim from Buchanan and Somers (1969), Rantz (1982), Simpson (2002), Oberg and others (2005), and Mueller and Wagner (2009), and even though some of these parts are not specifically denoted, credit is hereby given to these authors.

Definition of Streamflow

Streamflow, or discharge, is defined as the volumetric rate of flow of water (volume per unit time) in an open channel, including any sediment or other solids that may be dissolved or mixed with it that adhere to the Newtonian physics of open-channel hydraulics of water. The definition of streamflow in this chapter does not include non-Newtonian flow events such as debris flows and lahars (an avalanche of volcanic mud and water down the slopes of a volcano). Streamflow in the USGS is usually expressed in English dimensions of cubic feet per second (ft^3/s). Other common units are million gallons per day (Mgal/d) and acre-feet per day (ac-ft/d). Streamflow cannot be measured directly but must be computed from variables that can be measured directly, such as stream width, stream depth, and streamflow velocity. Even though streamflow is computed from measurements of other variables, the term "streamflow measurement" or "discharge measurement" is generally applied to the final result of the calculations.

Discharge Measurements at Gaging Stations

Procedures for making most types of current-meter [mechanical meters, electromagnetic meters, ADV meters, acoustic digital current meters (ADCs), and so forth)], moving-boat ADCP, and ADCP midsection measurements are described in the following sections. For much of the discussion of moving-boat ADCP, the reader is referenced to Mueller and Wagoner (2009). The chapter includes discussions on the selection of a measuring section, laying out the stationing for subsection verticals, width measurements, depth measurements, velocity measurements, direction of flow measurements, and recording of field notes. Additional details that pertain to instrumentation and specific types of measurements, such as wading, cableway, bridge, boat, and ice, are described in subsequent sections. Special procedures such as networks of current meters, measurement of deep, swift streams, and measurements during rapidly changing stage are also described.

Velocity-Area Method

The most practical method of measuring the discharge of a stream is the velocity-area method. Discharge is computed as the product of the area and velocity. The measurement is made by subdividing a stream cross section into segments (sometimes referred to as partial areas, sections, subareas, verticals, stations, profiles, panels, or ensembles), and by measuring the depth and velocity in a vertical within each segment. The total discharge is the summation of the products of the partial areas of the stream cross section and their respective average velocities. This computation is classically expressed by the equation

$$Q = \sum_{i=1}^{n} a_i v_i, \tag{1}$$

where Q total discharge, in cubic feet per second,
  a_i cross-section area, in square feet, for the ith segment of the n segments into which the cross section is divided, and
  v_i the corresponding mean velocity, in feet per second of the flow normal to the ith segment, or vertical.

Midsection Method

The current-meter midsection method of making a current-meter discharge measurement is used by the USGS and others. The method assumes that the mean velocity in each vertical represents the mean velocity in a partial rectangular area (segment). The mean velocity in each vertical is determined by measuring the velocity at one or more selected points in that vertical, as described in a later section of this chapter. The cross-section area for a segment extends laterally from half the distance from the preceding vertical to half

the distance to the next vertical, and vertically, from the water surface to the sounded depth as shown in figure 1.

The cross section in figure 1 is defined by depths at locations 1, 2, 3, 4, . . . , n. At each location, the velocities are sampled by current meter to obtain the mean of the vertical distribution of velocity. The partial discharge is now computed for any partial section (segment) at location i as

$$q_i = v_i \left[\frac{\left(b_i - b_{(i-1)}\right)}{2} + \frac{\left(b_{(i+1)} - b_i\right)}{2} \right] d_i, \text{ or} \tag{2}$$

$$= v_i \left[\frac{b_{(i+1)} - b_{(i-1)}}{2} \right] d_i, \tag{3}$$

where q_i discharge through partial section i,
  v_i mean velocity at location i,
  b_i distance from initial point to location i,
  $b_{(i-1)}$ distance from initial point to preceding location,
  $b_{(i+1)}$ distance from initial point to next location, and
  d_i depth of water at location i.

Thus, for example, the discharge through partial section 4 (heavily outlined in figure 1) is

$$q_4 = v_4 \left[\frac{b_5 - b_3}{2} \right] d_4. \tag{4}$$

The procedure is similar when i is at an end section. The "preceding location" at the beginning of the cross section is considered coincident with location 1; the "next location" at the end of the cross section is considered coincident with location n. Thus,

$$q_1 = v_1 \left[\frac{b_2 - b_1}{2} \right] d_1, \text{ and} \tag{5}$$

$$q_n = v_n \left[\frac{b_n - b_{(n-1)}}{2} \right] d_n. \tag{6}$$

For the example shown in figure 1, q_1 is zero because the depth at observation point 1 is zero. However, when the cross-section boundary is a vertical line at the edge of the water as at location n, the depth is not zero and velocity at the end section may or may not be zero. Equations 5 and 6 are used whenever there is water only on one side of an observation point, such as at the edge of the stream, piers, abutments, and islands. It usually is necessary to estimate the velocity at an end section because it normally is impossible to measure the velocity accurately with the current meter close to a boundary. There also is the possibility of damage to the equipment if the flow is turbulent. The estimated velocity is usually made as a percentage of the adjacent section.

The summation of the discharges for all the partial sections is the total discharge of the stream. An example of the measurement notes is shown in figure 2A. In the hydraulic properties reported, the summation of discharges from an ADV discharge

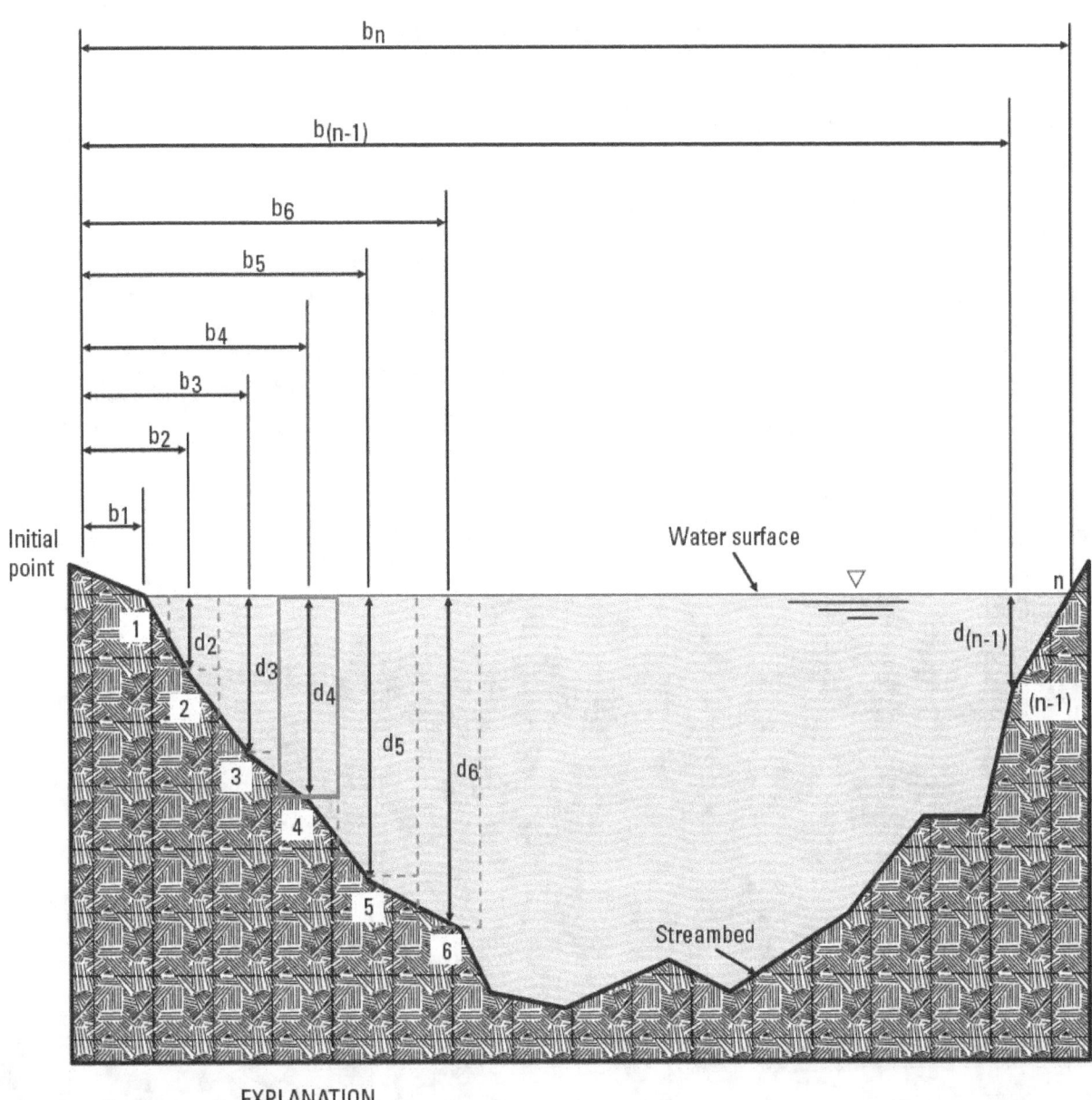

EXPLANATION

1, 2, 3,............... n	Observation points
$b_1, b_2, b_3, ... b_n$	Distance, in feet, from the initial point to the observation point
$d_1, d_2, d_3, ... d_n$	Depth of water, in feet, at the observation point
- - - - - - - - - - -	Boundary of partial sections; one heavily outlined discussed in text

Figure 1.   Definition sketch of the current-meter *midsection* method of computing cross-section area for discharge measurements.

measurement (fig. 2*B*) is similar to that of a current meter; however, it is designed to report the properties inherent to the ADV software and signal processing necessary to compute discharge using acoustic Doppler technology. A program written by staff in the USGS Maine Water Science Center entitled Surface Water Measurements and Inspections (SWAMI) has become common in use in the USGS with a PDA, and may be used to record discharge measurements, inspections, differential level surveys, and other field measurements (fig. 2*C*).

Included here for convenience is a typical, well-documented ADCP discharge measurement (fig. 2*D*). This measurement serves as an example of how an ADCP measurement note should be kept in the field. Further discussion of ADCP measurements can be found in subsequent sections of this chapter.

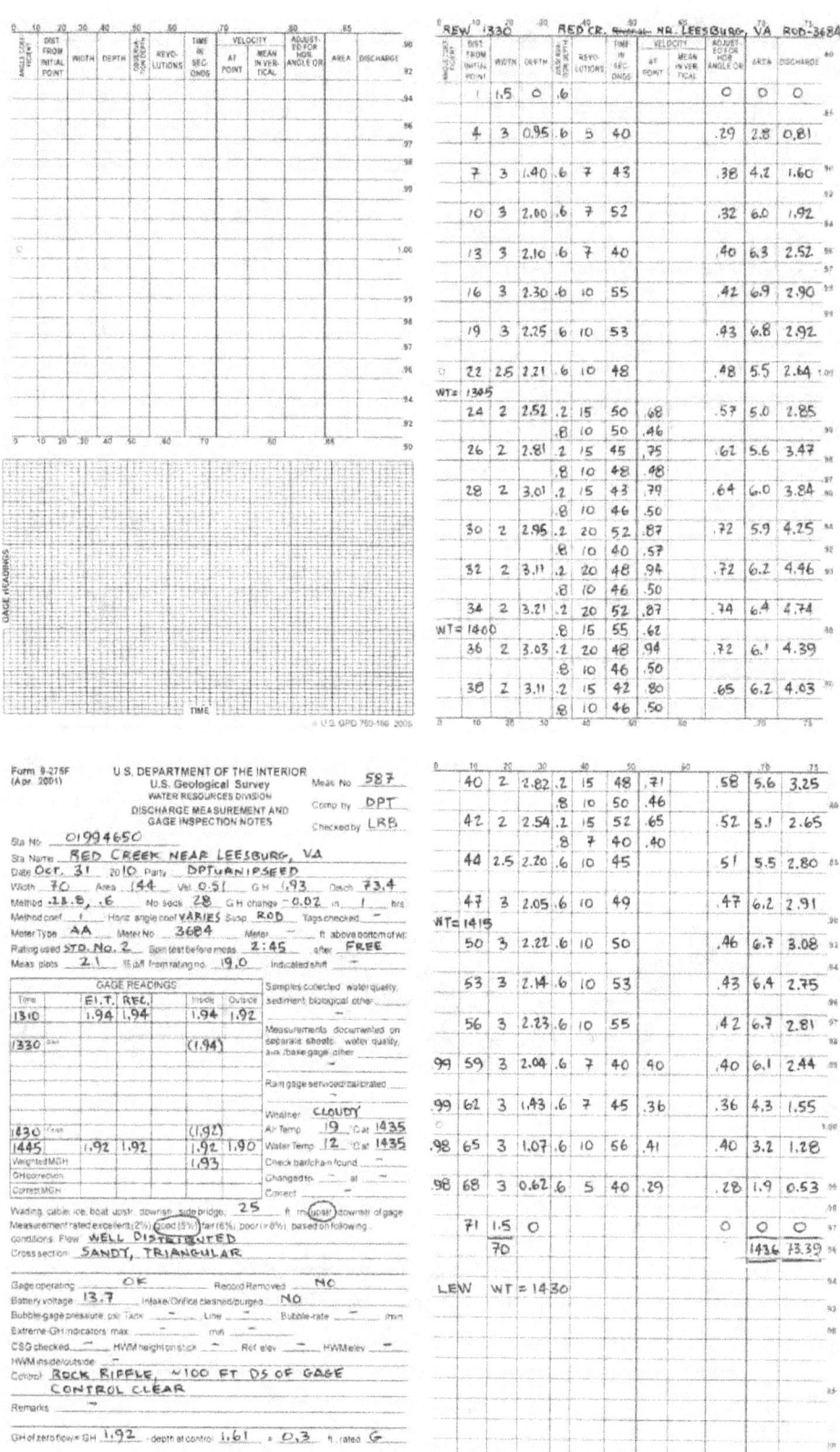

A

Figure 2. Computation notes of: *A*, a current-meter measurement by the *midsection* method; *B*, an ADV discharge measurement; *C*, from the Surface Water Measurement and Inspection (SWAMI) form; and *D*, an ADCP discharge measurement.

Discharge Measurement Summary

SEM
Date Generated: Tue Nov 17 2009

File Information

File Name	03329700.WAD
Start Date and Time	2009/08/18 11:23:50

Site Details
Site Name
Operator(s)   HTM

System Information

Sensor Type	FlowTracker
Serial #	P156A
CPU Firmware Version	3.3
Software Ver	2.20

Units (English Units)

Distance	ft
Velocity	ft/s
Area	ft^2
Discharge	cfs

Summary

Averaging Int.	40
Start Edge	LEW
Mean SNR	32.0 dB
Mean Temp	73.12 °F
Disch. Equation	Mid-Section

# Stations	30
Total Width	63.000
Total Area	51.550
Mean Depth	0.818
Mean Velocity	1.1593
Total Discharge	**59.7613**

Discharge Uncertainty

Category	ISO	Stats
Accuracy	1.0%	1.0%
Depth	0.2%	1.0%
Velocity	0.6%	3.1%
Width	0.1%	0.1%
Method	1.5%	
# Stations	1.7%	
Overall	**2.6%**	**3.4%**

Measurement Results

St	Clock	Loc	Method	%Dep	MeasD	Depth	Vel	CorrFact	MeanV	Area	Flow	%Q
0	11:23	0.00	None	0.0	0.0	0.000	0.0000	1.00	0.0000	0.000	0.0000	0.0
1	11:23	4.00	0.6	0.6	0.160	0.400	0.2966	1.00	0.2966	1.200	0.3559	0.6
2	11:25	6.00	0.6	0.6	0.200	0.500	0.4426	1.00	0.4426	1.000	0.4426	0.7
3	11:26	8.00	0.6	0.6	0.240	0.600	0.6266	1.00	0.6266	1.200	0.7521	1.3
4	11:27	10.00	0.6	0.6	0.320	0.800	0.7802	1.00	0.7802	1.600	1.2481	2.1
5	11:28	12.00	0.6	0.6	0.440	1.100	0.9803	1.00	0.9803	2.200	2.1568	3.6
6	11:29	14.00	0.6	0.6	0.520	1.300	0.8986	1.00	0.8986	2.600	2.3362	3.9
7	11:30	16.00	0.6	0.6	0.520	1.300	0.9829	1.00	0.9829	2.600	2.5554	4.3
8	11:32	18.00	0.6	0.6	0.520	1.300	1.1299	1.00	1.1299	2.600	2.9375	4.9
9	11:33	20.00	0.6	0.6	0.520	1.300	1.2979	1.00	1.2979	2.600	3.3742	5.6
10	11:34	22.00	0.6	0.6	0.440	1.100	0.9744	1.00	0.9744	2.200	2.1438	3.6
11	11:36	24.00	0.6	0.6	0.440	1.100	1.4452	1.00	1.4452	2.200	3.1797	5.3
12	11:37	26.00	0.6	0.6	0.480	1.200	0.9862	1.00	0.9862	2.400	2.3672	4.0
13	11:38	28.00	0.6	0.6	0.480	1.200	1.0472	1.00	1.0472	2.400	2.5137	4.2
14	11:39	30.00	0.6	0.6	0.400	1.000	0.9636	1.00	0.9636	2.000	1.9272	3.2
15	11:40	32.00	0.6	0.6	0.400	1.000	0.9551	1.00	0.9551	2.000	1.9101	3.2
16	11:41	34.00	0.6	0.6	0.400	1.000	1.5469	1.00	1.5469	2.000	3.0938	5.2
17	11:42	36.00	0.6	0.6	0.400	1.000	1.4954	1.00	1.4954	2.000	2.9906	5.0
18	11:44	38.00	0.6	0.6	0.400	1.000	1.4741	1.00	1.4741	2.000	2.9482	4.9
19	11:45	40.00	0.6	0.6	0.360	0.900	1.5633	1.00	1.5633	1.800	2.8138	4.7
20	11:46	42.00	0.6	0.6	0.360	0.900	1.3133	1.00	1.3133	1.800	2.3638	4.0
21	11:47	44.00	0.6	0.6	0.320	0.800	1.6240	1.00	1.6240	1.600	2.5980	4.3
22	11:48	46.00	0.6	0.6	0.320	0.800	1.2562	1.00	1.2562	1.600	2.0096	3.4
23	11:49	48.00	0.6	0.6	0.280	0.700	1.6437	1.00	1.6437	1.400	2.3016	3.9
24	11:51	50.00	0.6	0.6	0.280	0.700	1.4249	1.00	1.4249	1.400	1.9952	3.3
25	11:52	52.00	0.6	0.6	0.240	0.600	1.2969	1.00	1.2969	1.200	1.5565	2.6
26	11:53	54.00	0.6	0.6	0.240	0.600	1.4646	1.00	1.4646	1.200	1.7577	2.9
27	11:54	56.00	0.6	0.6	0.200	0.500	1.4452	1.00	1.4452	1.000	1.4452	2.4
28	11:55	58.00	0.6	0.6	0.200	0.500	0.9639	1.00	0.9639	1.750	1.6868	2.8
29	11:55	63.00	None	0.0	0.0	0.000	0.0000	1.00	0.0000	0.000	0.0000	0.0

Rows in italics indicate a QC warning. See the Quality Control page of this report for more information.

9-275-x   01/24/2007

U.S. DEPARTMENT OF THE INTERIOR
U.S. Geological Survey

Station Number	Meas. No.	625
03329700	Comp. by	SEM
	Checked by	RGK

ADV Discharge Measurement Notes

Station Name	Deer Creek nr Delphi, IN		
Date	8/18/2009	Party	SEMorlock

Width	Area	Velocity	SNR	Gage Height	Discharge
63	51.6	1.16	32	2.08	59.8

Method	# Sections	Gage Height Change
.6	30	-.01 in .52 hrs.

Std Velocity Profile   Y or N   Measured Water Temp 24.5°F/Cat 1100   Weather Cloudy
Diagnostic Test File P156.080509.bdc
Raw Data File P156.080509.WAD
03329700.WAD   ADV Sync'd to WT   Y at 1120 or N   ADV Water Temp 73.1 °F/ea at 1140   Wind Speed/Dir. 0-5/SW

Manufacturer	Model	Serial No.	Firmware	Software
Sontek	FlowTracker	P156A	3.3	2.20

EST   GTG STAFF

Time	Gage Reading			
	DCP	Inside	Outside	
1052		2.09	2.08	2.08
1123 S				
1155 F				
1207		2.08	2.07	2.07

Rating number	30
Percent from rating	3.5
Indicated shift	∅

Rain gage serviced/calibrated   n/a

Salinity   ___ ppt at ___

Checkbar found   no
Checkbar Changed to   ___ at ___

Weighted MGH
GH corrections
Correct MGH

Wading, cableway, boat, upstr, downstr, side bridge   ft, mi, upstr, downstr, of gage   based on following conditions
Measurement rated   excellent (2%), good (5%), fair (8%), poor (>8%)
Flow   Steady and Even
Cross section   Sand and gravel, fairly even
Control   Sand, gravel and cobbles 100'ds clear

Gage operating   Y or N   Record removed   Y or N   Intakes/orifice cleaned/purged   Yes   Filename 03329700.109

Battery voltage	12.8 v	Tank	na
Bubble-gage psi		Line	na
Extreme-GH indicators		Max	2.45
HWM on stick	no		
GH of zero flow = GH	2.08	Ref elev.	4.32

Bubble rate   1.87   / min
Min   ___
CSG Checked   Y or N   CSG elevation   na
HWM elevation na
- depth at control   1.20   = .88 ft.   Rated Fair

Remarks

Sheet No. 1   of 2   sheets

B

C

Acoustic Profiler Discharge Measurement Notes

Filename Prefix: foxmon_ds1200_

Left Bank: (Sloping) Vertical Other:_____   Right Bank: (Sloping) Vertical Other:_____

Transect No.	Bank	Starting Time	Starting Distance	Ending Distance	Ending Time	Total Discharge	Notes
0	L R						Moving bed test in center of channel
1	L (R)	1249	16	69	1255	1,321	Simultaneous comparison discharge measurements
2	(L) R	1256	69	16	1301	1,358	upstream of dam
3	L (R)	1301	69	———			Transect aborted due to debris in river
4	L (R)	1303	69	16	1308	1,327	
5	(L) R	1309	16	69	1315	1,356	
	L R						
	L R						
	L R						
	L R						
	L R						
	L R						
	L R						

Notes: All times are in CST. Measurement was made using a temporary rope-and-pulley cableway. Edge distances were measured with laser rangefinder by marking the start and ending positions on the rope and measuring the distance from edge of water to the center of the tethered boat.

U.S. DEPARTMENT OF THE INTERIOR
U.S. Geological Survey

ADCP Discharge Measurement Notes

9-275-J 7/19/08

Meas. No.	57
Processed by	BLL
Checked by	KAO

Station Number	05551540
Station Name	Fox River at Montgomery, IL
Date	July 6, 2004
Party	B.L. Loving, S.E. Anderson

Width	235	Area / Rated Area	707	Velocity	1.90	Index Vel		Gage Height	11.74	Discharge	1,340

Gage Height Change	0.00 in 0.4 hrs	From rating	No.: 11	Meas. plots	4.4% diff	Shift	0.0	ADCP Sync'd to WT	(Y) at 1207 or N

ADCP Mfr / Model / Frequency	RDI Rio Grande 1200	Serial No.	1636	Firmware		Software	WinRiver 10.06

Boat/Motors Used	OceanScience Tethered	GPS Used	Trimble AgGPS	ADCP Depth	0.27 ft	Gage Height	10.14	Diag. Test / Errors?	Y or (N)

Compass Calib & Total Error	0.8	Mag Var	-2.4	MagVar Method	On-site (Model) Previous	Moving Bed?	(Y) or N

Meas. Water Temp	26.5°F @ 1210	ADCP Water Temp	26.5°F @ 1210	Weather / Air Temp	Sunny/clear 85(F)°C	Wind Speed / Dir	Southerly @ 5-10 mph

Site Conditions

Max Water Depth	10 ft	
Max Water Speed	2.5 ft/s	
Max Boat Speed	1 ft/s	
Water Mode	12	
Bottom Mode	5	
Streambed material	Gravel	
Salinity		ppt at
Checkbar found	22.41	
Checkbar changed to:	— at —	

Gage Readings

Time	ETG	CR10	Inside	Outside		
1100	11.74	11.70		11.70		
1230		11.70		±0.05		
1249		11.70			(S)	
1300		11.70				
1315		11.70			(F)	
1400	11.74	11.70		11.70		
				±0.05		
Weighed MGH	11.74					
GH corrections						
Correct MGH	11.74					

Wading, cable, ice, boat, upstr., downstr., side bridge 1500 (ft) mi upstr. (downstr.) of gage

Measurement rated: excellent (2%), (good (5%)) fair (8%), poor (>8%) based on following conditions

Flow: Steady & uniform. Flow at edges appears to be moving in DS direction

Cross section: Sand and Gravel with some mud

Control: Dam is clear of debris

Gage operating:	(Y) or N	Record removed	Y or (N)	Filename		Telephone telemetry
Battery voltage	12.5 V	Intakes/Orifice cleaned/purged:	No			
Bubble-gage psi	Tank —	Line —				— / min
Extreme-GH indicators	Max —	Min —			CSG Checked	(Y) or N
HWM on stick	None	Ref elev:	12.65	HWM elevation:	None	Rated= 10.48
GH of zero flow = GH	—	- depth at control	—	Sheet No.	1	of 1 — ft. sheets

D

The mean-section method used by the USGS prior to 1950 differs from the midsection method in computation procedure. Partial discharges are computed for partial sections between successive verticals. The velocities and depths at successive verticals are each averaged, and each partial section extends laterally from one vertical to the next. Discharge is the product of the average of two mean velocities, the average of two depths, and the distance between verticals. A study by Young (1950) concluded that the midsection method is simpler to compute and is a slightly more accurate procedure than the mean-section method.

Site Selection

The first and most critical step in making a midsection current-meter or ADV measurement, or an ADCP measurement is to select a measurement cross section of desirable qualities. If the stream cannot be waded, nor high-water measurements made from a bridge, moving or tethered boat, or cableway, the hydrographer may have little or no choice in selecting a measurement cross section. If the stream can be waded or the measurement can be made from a boat, the hydrographer should look for a cross section with the following characteristics:

- There is a reasonably straight channel with streamlines parallel to each other; a stable streambed free of large rocks, weeds, and obstructions that would create eddies, slack water, and turbulence; and desirable measurement sections that are roughly parabolic, trapezoidal, or rectangular. These conditions are obviously not always possible, but remember that most current meters are rated in a still water tank by towing them through the tank at a known speed. With that in mind, these are conditions a hydrographer should seek in the field: a smooth, mirror-like water surface with steady, uniform, nonvarying flow conditions in the stream reach where the discharge measurement will be taken.

- The velocities are, for the most part, greater than 0.5 ft/s, and depths that are greater than about 0.5 ft. These conditions are not always possible to find in the field.

- The measurement section is relatively close to the gaging station control to avoid the effect of tributary and (or) intervening drainage area inflows between the measurement section and the control, and to avoid the effect of channel storage between the measurement section and the control during periods of changing stage.

It is usually not possible to satisfy all of these conditions. Select the best possible reach using these criteria and then select a cross section. For a further discussion regarding site selection when using a mechanical or other point-velocity current meter refer to Rantz and others (1982).

For convenience, special site-selection considerations for an ADCP discharge measurement are presented as follows, and further discussion of ADCP methods and instruments is presented in subsequent sections of this chapter:

- The minimum depth near the left and right edges of water at the measurement site should allow for the measurement of velocity in two or more depth cells while being close enough to minimize the estimated edge discharges.

- Make sure velocities are, for the most part, greater than 0.5 ft/s, and depths are greater than the minimum depth required by the ADCP. Although measurements can be made in low velocities, keep boat speeds extremely slow (if possible, less than or equal to the average water velocity), which requires special techniques for boat control (Simpson, 2002).

- Avoid measurement sections having local magnetic fields, especially if a moving bed is present and a Global Positioning System with differential corrections (DGPS) or the Loop Method (Mueller and Wagoner, 2006) is used. For example, during measuring, avoid overhead truss bridges, low steel-beam spans, power lines, and other sources of magnetic fields. Just as with ADCP mounts and boats, the presence of ferrous metals will result in ADCP compass errors.

- If possible, avoid asymmetric channel geometries (for example, deep on one side and shallow on the other; Simpson, 2002) and avoid cross sections with abrupt changes in channel-bottom slope. The streambed cross section should be as uniform as possible and free from debris and vegetation or plant growth.

- When using DGPS with an ADCP, avoid cross-section locations where multipath interference, such as riparian vegetation (low-hanging trees and large bushes on river or stream banks), buildings at or near the river banks, bridges, and other flow-control structures, could impede or block signals from GPS satellites.

It is usually not possible to attain all of these conditions, but site selection cannot be understated as a critical part of a discharge measurement. Select the best possible reach using these criteria and then select a cross section. For more discussion regarding site selection when using an ADCP, refer to Mueller and Wagner (2009).

Layout and Stationing of Partial Sections and Verticals in a Midsection Current-Meter Discharge Measurement

After the cross section has been selected, determine the width of the stream. For a mechanical current-meter or other point-velocity measurement, string a tag line or measuring tape for measurements made by wading, from a boat, from ice cover, or from an unmarked bridge. Except for bridges, string the line at right angles to the direction of flow to avoid horizontal angles in the cross section. For cableway or bridge measurements, use the graduations painted on the cable or bridge rail. Next, determine the spacing of the verticals, generally using about 25 to 30 partial sections. With a smooth cross section and even velocity distribution, fewer partial sections may be used. Space the partial sections so that no partial section has more than 10 percent of the total discharge in it. The ideal measurement is one in which no partial section has more than 5 percent of the total discharge in it; this can be challenging when only 25 partial sections are used. For example, the discharge measurement shown in figure 2A had 6.5 percent of the total discharge in the partial section with the greatest discharge. Equal widths of partial sections across the entire cross section are not recommended unless the discharge is evenly distributed. Lessen the width of the partial sections as depths and velocities become greater. Usually an approximate or expected total discharge can be obtained from the stage-discharge curve. Space the verticals so the discharge in each partial section is about 5 percent of the expected total discharge from the rating curve. When using an electronic field notebook [such as the JBS Instruments Aquacalc Pro Discharge Measurement Computer (Aquacalc), a PDA with the Hydrological Services Current Meter Counter signal processor (CMCsp), or the SonTek FlowTracker], the expected total discharge can be entered prior to starting the discharge measurement. During the measurement, a warning message will be displayed if a partial discharge exceeds 10 percent of the expected total discharge. When using an ADV or other acoustic point-velocity instrument, make sure the instrument is appropriately aligned and plumbed to the tag line because slight variations in the alignment of the instrument can result in large errors in the measurement of point velocity. See further discussion of the use of acoustic point-velocity instruments in this chapter.

For a standard mechanical current-meter discharge measurement, the usual procedure, after selecting and laying out the section, is to measure and record at each vertical (1) the distance from the initial point, (2) the depth, (3) the meter position, (4) the number of revolutions, (5) the time interval, and (6) the horizontal angle of flow. The starting point can be either bank. The edge of water, which may have a depth of zero, is considered to be the first vertical. The hydrographer should move to each of the verticals in succession and repeat the procedure until the measurement is completed at the opposite bank.

Measurement of Width

The first measurement made in a discharge measurement is usually the determination of horizontal stationing (width) in the cross section being measured. Width needs to be measured using the proper equipment and procedures that apply to the type of measurement being made (that is, wading, bridge, cableway, boat, or ice). Details of measuring width using a variety of equipment, and under different flow conditions, are described in subsequent sections of this chapter.

The horizontal distance to any vertical in a cross section is measured from an initial point on the bank. Cableways and bridges used regularly for making discharge measurements are commonly marked at 2-, 5-, 10-, and (or) 20-ft intervals by paint marks. Distance between markings is interpolated, or measured with a rule or pocket tape. Steel or Kevlar tag lines and metallic tapes are used for measurements made by wading, from boats, or from unmarked bridges. For wide streams of about 2,500 ft or more, where conventional measuring methods cannot be used, surveying methods and Global Positioning Systems (GPS) can be used.

Tapes and Tag Lines

Tag lines used for wading measurements are usually made of either galvanized steel aircraft cord with solder beads at measured intervals, or Kevlar, which is marked with black ink and waxed to resist abrasion. A Kevlar tag line consists of a Kevlar core with a nylon jacket.

The standard arrangement of solder beads on steel tag lines is shown in table 1. The standard markings for Kevlar tag lines is one mark every 2 ft, two marks every 10 ft, and three marks every 100 ft. The standard lengths of tag lines are 300, 400, and 500 ft, but other sizes are available.

Four types of tag-line reels typically used for the steel tag lines are the Lee-Au, Pakron, Columbus type A, and the USGS Stainless Steel Tag line as shown in figure 3. The reel used for the Kevlar tag line is shown in figure 4.

Larger reels, used for boat measurements, are designed to hold up to 3,000 ft of ⅛-inch (in.) diameter steel tag line. These reels and boat measurement methods have largely been replaced by the ADCP technology. Two different types of reels still available are as follows:

- A heavy-duty, horizontal-axis reel without a brake, and with a capacity of 5,000 ft of ⅛-in. beaded tag line or 3,000 ft of $^3/_{16}$-in. Kevlar boat tag line, as shown in figure 5.

- A vertical-axis reel without a brake (fig. 6), and with a capacity of 1,500 ft of ⅛-in diameter steel tag line (800 ft tag lines are standard) or up to 900 ft of 3/16-in. Kevlar boat tag-line cable.

Table 1. Standard markings for steel tag lines.

Distance from initial point (zero mark), in feet	Distance between marks, in feet	Number of solder beads, or tags
0 to 50	2	1 (single bead)
50 to 100	5	1
150 to 500	10	1
0 to 50	10	2 (double bead)
50 to 450	100	2
0 to 500	100	3 (triple bead)

Figure 3. Type-A reels: *A*, Lee-Au; *B*, Pakron; *C*, Columbus; and *D*, USGS stainless steel tag line.

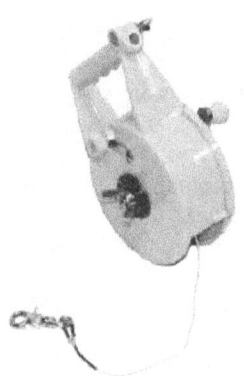

Figure 4. Kevlar tag-line reel.

Figure 5. Horizontal-axis boat tag-line reel without a brake.

of a surveying level, as shown in figure 7, or electronic total station instrumentation, as shown in figure 8. The procedure to determine width with a transit is used less frequently, but is described in the section of this chapter on boat measurements.

With the advent of electronic total station surveying instruments and digital surveying levels, a direct reading of the distance can be made from the digital surveying level or from the total station instrument setup point to the boat. One example of a commercially available total station instrument is shown in figure 8. Most of these instruments require a reflector target at the point where a measurement is desired (in this case the boat); however, total station instruments are also available that provide accurate measurements of distance without a reflector target. Accurate distance measurements can be made with total station instruments over distances of 1 mi or more, provided the boat can be seen and not obstructed by intervening objects.

Figure 6. Vertical-axis boat tag-line reel.

Surveying Methods of Width Measurement, Surveying Level, and Electronic Total Station

For wide streams where it is not practical to string a tag line for discharge measurements from a boat, surveying methods can be used to measure stream width and stationing for measurement points. Surveying methods require the use

Figure 7. Surveying level and tripod.

Figure 8.   Total station surveying equipment: A, instrument and tripod and B, instrument reflector targets and surveying tape.

Global Positioning System With Differential Corrections

Measurement points can be stationed for wide streams, such as flood plains that may be several miles wide, or large tidal estuaries, with a global positioning system with differential corrections (DGPS) instrument, such as that shown in figure 9, which is a real-time kinematic (RTK) GPS base station with radio transmitter. RTK-GPS and other DGPS instruments utilize satellite telemetry from an array of satellites, and use radio triangulation to compute positions for any point on the Earth. In order to obtain the accuracy necessary for a discharge measurement, the raw GPS positions must have differential corrections applied on the basis of simultaneous readings at a base station. Most DGPS units contain built-in differential correction receivers that make automatic and instantaneous corrections. Other GPS units may use a separate receiver that attaches to the GPS unit with a cable. In either case, base station data are received by radio signal from nearby ground base stations or via communications satellites from a network of ground stations, such as the Wide Area Augmentation System (WAAS).

A surveying grade DGPS unit with capability of storing and recalling data is preferred. These units may or may not have built-in or attached differential correction receivers. If instantaneous differential corrections are not made, then use correction data obtained after the fact from a separate GPS base station to postprocess coordinate data. Various agencies collect and provide the base station data via the Internet. Coordinate data for the measurement points are downloaded from the GPS unit to a computer for postprocessing. Software is available to make the differential corrections, to compute corrected coordinates of the measurement points, and to automatically compute distances between and to plot a map of the measurement points.

Figure 9.   Global positioning system with real-time kinematic (RTK–GPS) base station and radio transmitter.

Accuracy of GPS coordinates will vary depending on the type of GPS unit used and whether or not differential corrections are made. Coordinates without differential corrections can be in error by as much as ±300 ft because of various errors in the system. Obviously, this is not acceptable for discharge measurements. However, if care is taken in making observations, and then making differential corrections, errors can be reduced to as little as ±3 ft, and even less in ideal conditions. This method is acceptable for wide flood plains and inaccessible estuaries with open skies and minimal reflective surfaces, which can result in multipath errors.

Measurement of Depth

The second measurement normally made at a vertical is the stream depth. Depth should be measured using the proper equipment and procedures that apply to the type of measurement being made (that is, wading, bridge, cableway, boat, or ice). Details of measuring depth using various equipment and under different flow conditions are described in the following sections of this chapter. The water depth of a stream at a selected vertical can be measured in several ways, depending on the type of measurement being made, the total depth of the stream, and the velocity of the stream. Stream depth is usually measured by use of a wading rod, sounding lines and weights, acoustic Doppler sensor, or another sonic sounder, as described in the following sections of this chapter.

Use of Wading Rod

Use a wading rod for measuring stream depth when depth is shallow enough, or when measuring from a low footbridge or other supportive structure over the stream. Likewise, use the wading rod for measuring from ice cover for shallow depths. Wading rods can even be used from a boat if depths are not too great. The top-setting wading rod can be used for depths up to 4 ft, but greater depths can be measured with 6-, 8-, and 10-ft top-setting wading rods. The round wading rod, which is assembled with 1-ft sections, can be made up into any length, but generally is not used for depths greater than about 10 ft. Velocity of flow is also a consideration because high velocity may not allow for keeping a long wading rod in place.

Wading rods have a small foot on the bottom to allow the rod to be placed firmly on the streambed, and yet not sink into the streambed under most conditions. In sand-bottom streams, or in soft muck, it is sometimes difficult to keep the wading rod from sinking into the streambed as the weight of the rod and meter and the eroding power of the flowing water cause the foot of the wading rod to sink. The hydrographer must use care in these conditions to be sure the measured water depth, as well as the depth of the current-meter placements, are accurately based on the surface of the streambed. In some cases, the wading rod may need to be supported in some manner other than resting on the streambed.

When using a wading rod in streams with moderate-to-high velocity, there will be a velocity-head build-up of water on the wading rod. The stream depth should be based on where the surface of the stream intersects the wading rod, and not on the top of the velocity-head build-up. Wading rods are graduated in tenths-of-a-foot, and stream depths are generally measured or estimated and recorded to the nearest 0.01 ft.

Use of Sounding Lines and Weights

Water depth is measured with sounding lines and weights when the depth is too great to use a wading rod, and when measuring conditions require measuring from a bridge, cableway, or boat. This section will describe the measurement of depth when using sounding reels and handlines. It also discusses the procedures used to correct observed depths when high velocity causes the weight and meter to drift downstream.

Use of Sounding Reels

When using one of the sounding reels described in a subsequent section of this chapter, a counter or dial is used to determine the length of cable that has been dispensed. Depths are measured to the nearest 0.1 ft when using a sounding line and weight.

The size of the sounding weight used in current-meter measurements depends on the maximum depth and velocity in a cross section. A rule of thumb is that the size of the weight in pounds should be greater than the maximum product of velocity and depth in the cross section. If insufficient weight is used, the sounding line will be dragged at an angle downstream. If debris or ice is flowing or if the stream is shallow and swift, a heavier weight can be used than the rule designates. The rule is not rigid but it does provide a starting point for deciding on the size of the weight that is needed. If available, notes can be examined of previous measurements at a site to help determine the size of the weight needed at various stages.

Some sounding reels are equipped with a computing depth indicator, or spiral. To use the computing spiral, the dial pointer must be set at zero when the center of the current-meter rotor is at the water surface. After the sounding weight and meter are lowered until the weight touches the streambed, and the indicated depth should be read. The distance that the meter is mounted above the bottom of the weight should be added. For example, if a 30 C .5 (that is, a 30-pound Columbus weight is being used and the center of the meter cups is 0.5 ft above the bottom of the weight) suspension is used and the dial pointer reads 18.5 ft when the sounding weight touches the streambed, the depth would be 19.0 ft (18.5 + 0.5). To move the meter to the 0.8-depth position, merely raise the weight and the meter until the pointer is at the 19-ft mark on the graduated spiral, which will correspond to 15.2 ft on the main dial (0.8 × 19.0). To set the meter at the 0.2-depth position, raise the weight and meter until the pointer is at 3.8 ft on the main dial (0.2 × 19.0).

Tags can be placed on the sounding line a known distance above the center of the meter cups as an aid in determining depth. The tags, which are usually streamers of different-colored binding tape, are fastened to the sounding line by solder beads or by small cable clips. Tags are used for determining depth in two ways; the following is the preferred procedure:

1. Set the tag at the water surface and then set the depth indicator to read the distance of that tag above the center of the meter cups.

2. Continue as if the meter cups themselves have been set at the water surface.

3. When the weight touches the streambed, read the depth indicator and add the distance that the meter is above the bottom of the weight to obtain the total depth.

4. Use the spiral indicator, as described above, for setting the 0.8-depth meter position. If debris or ice is flowing, this method keeps the meter below the water surface and helps to prevent damage to the meter.

This is an alternate method that is sometimes used with handlines and sounding reels: With the sounding weight on the streambed, raise the weight until the first tag below the water surface appears at the surface. If using a reel, determine the distance the weight was raised by subtracting before and after readings of the depth indicator; if using a handline, use a tape or measuring stick. The total stream depth is the sum of (a) the distance the weight was raised to bring the tag to the water surface, (b) the distance the tag is above the center of the meter cups, and (c) the distance from the bottom of the weight to the center of the cups.

Use of a Handline

Although rarely used in the USGS, handlines still provide a viable means of measuring discharge from bridges. When using a handline, unwind enough cable from the handline reel to keep the reel out of water when the sounding weight is on the streambed at the deepest part of the cross section. If the bridge is high enough above the water surface, raise and lower the weight and meter by the rubber-covered cable rather than by the bare cable.

The usual procedure for determining depths is to set the meter cups at the water surface and then lower the sounding weight to the streambed while measuring the amount of line needed to reach the streambed. Measure along the rubber-covered service cord with a steel or metallic tape or a graduated rod to determine the distance the weight is lowered. This measured distance, plus the distance from the bottom of the sounding weight to the meter cups, is the depth of water. When the meter is set for the velocity observation, stand on the rubber-covered cable or tie it to the handrail to hold the meter in place. This arrangement frees the hands to record the data.

Another method of determining depth when using a handline includes the use of tags set at a known distance above the meter. Lower the sounding weight to the streambed, and then raise the weight until one of the tags is at the water surface. Measure along the rubber-covered service cord with a steel or metallic tape or a graduated rod to determine the distance the weight is raised. The total depth of water is then the summation of (1) the distance the particular tag is above the meter cups, (2) the measured distance the meter and weight was raised, and (3) the distance from the bottom of the weight to the meter cups.

Depth Corrections for Downstream Drift of Current Meter and Weight

Where it is possible to sound but the weight and meter drift downstream, the depths measured by the usual methods are too great. Figure 10 graphically illustrates this condition. The correction for this error has two parts, the air correction and the wet-line correction. The air correction is shown in figure 10 as the distance cd. The wet-line correction in figure 10 is shown as the difference between the wet-line depth de and the vertical depth dg.

As shown in figure 10, the air correction depends on the vertical angle P and the distance ab. The correction is computed as follows:

$$\cos P = \frac{ab}{ad} = \frac{ab}{ac+cd} = \frac{ab}{ab+cd} \Rightarrow$$
$$ab + cb = \frac{ab}{\cos P} \Rightarrow$$
$$cd = \frac{ab}{\cos P} - ab = ab\left[\frac{1}{\cos P} - 1\right]$$

(7)

where $\quad ab = ac$

The air correction for even-numbered angles between 4 degrees and 36 degrees and vertical lengths between 10 and 100 ft is shown in table 2. The correction is applied to the nearest tenth of a foot; hundredths are given to aid in interpolation.

Use of an air correction table may be nearly eliminated by using tags at selected intervals on the sounding line and using the tags to refer to the water surface. This practice is almost equivalent to moving the reel to a position just above the water surface.

The correction for excess length of line below the water surface is obtained by using an elementary principle of mechanics. If a known horizontal force is applied to a weight suspended on a cord, the cord takes a position of rest at some angle with the vertical. The tangent of the vertical angle of the cord is equal to the horizontal force divided by the vertical force owing to the weight. If several additional horizontal and vertical forces are applied to the cord, the tangent of the angle in the cord above any point is equal to a summation of the horizontal forces below that point, divided by the summation of the vertical forces below the point.

The distribution of total horizontal drag on the sounding line is in accordance with the variation of velocity with depth. The excess in length of the curved line over the vertical depth is the sum of the products of each tenth of depth and the function $[(1/\cos P) - 1]$ of the corresponding angles. The function is derived for each tenth of depth by means of the tangent relation of the forces acting below any point.

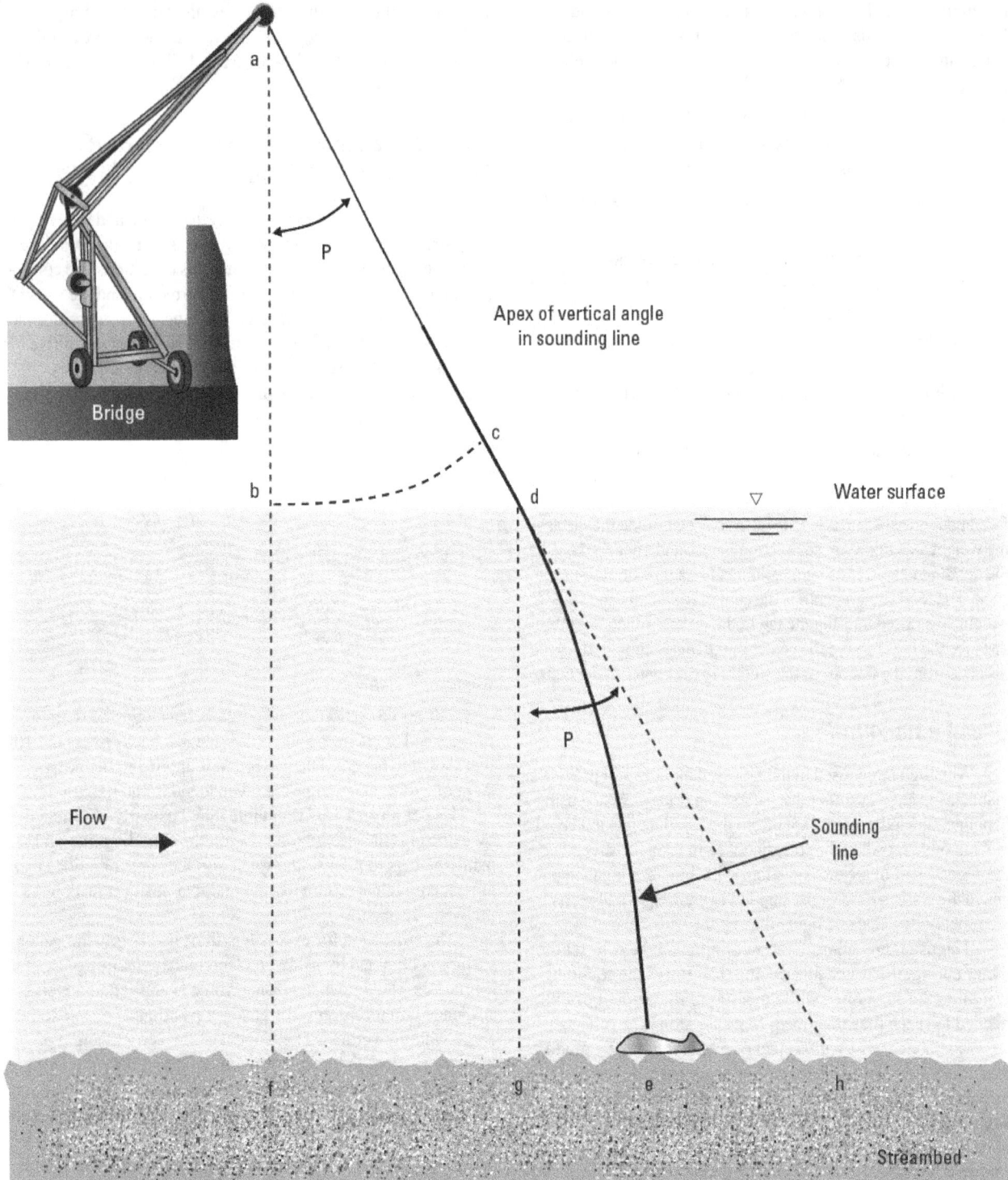

Figure 10.   Position of sounding weight and line in deep, swift water.

The wet-line correction for even-numbered angles between 4 degrees and 36 degrees and wet-line depths between 10 and 100 ft is shown in table 3. The correction is applied to the nearest tenth of a foot. The wet-line correction cannot be determined until the air correction has been deducted from the observed depth.

The following assumptions were used in deriving the wet-line correction table:

1. The weight will go to the bottom despite the force of the current.

2. The sounding is made when the weight is at the bottom but entirely supported by the line.

3. Drag on the streamlined weight in the sounding position is neglected.

4. The air-line/wet-line table is generic and can be used for any size sounding weight or line, provided they are designed to offer little resistance to the current.

Wet-line corrections can also be computed with the equation below. This polynomial equation was derived by Kenneth L. Wahl (U.S. Geological Survey [ret.], Regional Surface Water Specialist, written commun., 2000) from the data in table 3, and reproduces the table values to within a few hundredths of a foot. It can be used in a field computer to quickly and easily compute wet-line corrections. With additional programming, depths and depth settings can also be computed.

$$CORR_{wl} = D_{wl} - D_v = \left(0.0004081 - 0.0001471 \times P + 0.00005731 \times P^2\right) \times D_{wl}, \qquad (8)$$

where $CORR_{wl}$ the correction, in feet, to subtract from the wet-line depth to obtain the vertical depth,

D_{wl} the wet-line depth, in feet,

D_v the vertical depth, in feet, and

P the vertical angle, in degrees.

If the direction of flow is not perpendicular to the measuring section, the observed angle in the measuring line as indicated by a protractor will be less than the true angle of the line. The air correction and wet-line correction will then be too small. To correct for this, measure the horizontal angle between the direction of flow and a perpendicular to the measuring section with a protractor, or determine the horizontal angle coefficient as described in a subsequent section of this publication. The geometry of this condition is illustrated in figure 11.

If the horizontal angle of the direction of flow is called H, the observed vertical angle P, and the true vertical angle X, the relation between the angles is expressed by the equation

$$\tan X = \frac{\tan P}{\cos H}. \qquad (9)$$

Table 4 gives the amounts in tenths of degrees. Add these to observed vertical angles to obtain the true vertical angles for a range of horizontal angles between 8 degrees and 28 degrees.

The conditions that cause error in sounding the depth also cause error in the placement of the meter at selected depths. The correction tables are not strictly applicable to the problem of placing the meter because of the increased pressure placed on the sounding weight by higher velocities when it is raised from the streambed. A meter placed in deep, swift water by the ordinary methods for observations at selected percentages of the depth will be too high in the water. The use of tables 2 and 3 will tend to eliminate this error in placement of the meter, and although not strictly applicable, their use for this purpose has become general.

For the 0.2-depth position, the curvature of the wet line is assumed to be negligible and the length of sounding line from the apex of the vertical angle to the weight is considered to be a straight line. The method used to place the meter at the 0.2-depth position is as follows:

1. Compute the 0.2 value of the vertical depth.

2. Lower the meter to this depth into the water and read the vertical angle.

3. Obtain the air correction from table 2. The vertical length used to obtain the air correction is the sum of (a) 0.2 of the vertical depth, (b) the distance from the water surface to the apex of the angle, and (c) the distance from the bottom of the weight to the meter.

4. Let out an additional amount of line equal to the air correction.

Table 2. Air correction table (in feet), giving differences between vertical length and slant length of sounding line above water surface for selected vertical angles.

Vertical length (feet)	Vertical angle of sounding line at protractor																	Vertical length (feet)
	4°	6°	8°	10°	12°	14°	16°	18°	20°	22°	24°	26°	28°	30°	32°	34°	36°	
10	0.02	0.06	0.10	0.15	0.22	0.31	0.40	0.51	0.64	0.79	0.95	1.13	1.33	1.55	1.79	2.06	2.36	10
12	0.03	0.07	0.12	0.19	0.27	0.37	0.48	0.62	0.77	0.94	1.14	1.35	1.59	1.86	2.15	2.47	2.83	12
14	0.03	0.08	0.14	0.22	0.31	0.43	0.56	0.72	0.90	1.10	1.32	1.58	1.86	2.17	2.51	2.89	3.30	14
16	0.04	0.09	0.16	0.25	0.36	0.49	0.64	0.82	1.03	1.26	1.51	1.80	2.12	2.48	2.87	3.30	3.78	16
18	0.04	0.10	0.18	0.28	0.40	0.55	0.73	0.93	1.16	1.41	1.70	2.03	2.39	2.78	3.23	3.71	4.25	18
20	0.05	0.11	0.20	0.31	0.45	0.61	0.81	1.03	1.28	1.57	1.89	2.25	2.65	3.09	3.58	4.12	4.72	20
22	0.05	0.12	0.22	0.34	0.49	0.67	0.89	1.13	1.41	1.73	2.08	2.48	2.92	3.40	3.94	4.54	5.19	22
24	0.06	0.13	0.24	0.37	0.54	0.73	0.97	1.24	1.54	1.88	2.27	2.70	3.18	3.71	4.30	4.95	5.67	24
26	0.06	0.14	0.26	0.40	0.58	0.80	1.05	1.34	1.67	2.04	2.46	2.93	3.45	4.02	4.66	5.36	6.14	26
28	0.07	0.15	0.28	0.43	0.63	0.86	1.13	1.44	1.80	2.20	2.65	3.15	3.71	4.33	5.02	5.77	6.61	28
30	0.07	0.17	0.29	0.46	0.67	0.92	1.21	1.54	1.93	2.36	2.84	3.38	3.98	4.64	5.38	6.19	7.08	30
32	0.08	0.18	0.31	0.49	0.71	0.98	1.29	1.65	2.05	2.51	3.03	3.60	4.24	4.95	5.73	6.60	7.55	32
34	0.08	0.19	0.33	0.52	0.76	1.04	1.37	1.75	2.18	2.67	3.22	3.83	4.51	5.26	6.09	7.01	8.03	34
36	0.09	0.20	0.35	0.56	0.80	1.10	1.45	1.85	2.31	2.83	3.41	4.05	4.77	5.57	6.45	7.42	8.50	36
38	0.09	0.21	0.37	0.59	0.85	1.16	1.53	1.96	2.44	2.98	3.60	4.28	5.04	5.88	6.81	7.84	8.97	38
40	0.10	0.22	0.39	0.62	0.89	1.22	1.61	2.06	2.57	3.14	3.79	4.50	5.30	6.19	7.17	8.25	9.44	40
42	0.10	0.23	0.41	0.65	0.94	1.29	1.69	2.16	2.70	3.30	3.97	4.73	5.57	6.50	7.53	8.66	9.91	42
44	0.11	0.24	0.43	0.68	0.98	1.35	1.77	2.26	2.82	3.46	4.16	4.95	5.83	6.81	7.88	9.07	10.39	44
46	0.11	0.25	0.45	0.71	1.03	1.41	1.85	2.37	2.95	3.61	4.35	5.18	6.10	7.12	8.24	9.49	10.86	46
48	0.12	0.26	0.47	0.74	1.07	1.47	1.93	2.47	3.08	3.77	4.54	5.40	6.36	7.43	8.60	9.90	11.33	48
50	0.12	0.28	0.49	0.77	1.12	1.53	2.02	2.57	3.21	3.93	4.73	5.63	6.63	7.74	8.96	10.31	11.80	50
52	0.13	0.29	0.51	0.80	1.16	1.59	2.10	2.68	3.34	4.08	4.92	5.86	6.89	8.04	9.32	10.72	12.28	52
54	0.13	0.30	0.53	0.83	1.21	1.65	2.18	2.78	3.47	4.24	5.11	6.08	7.16	8.35	9.68	11.14	12.75	54
56	0.14	0.31	0.55	0.86	1.25	1.71	2.26	2.88	3.59	4.40	5.30	6.31	7.42	8.66	10.03	11.55	13.22	56
58	0.14	0.32	0.57	0.89	1.30	1.78	2.34	2.98	3.72	4.55	5.49	6.53	7.69	8.97	10.39	11.96	13.69	58
60	0.15	0.33	0.59	0.93	1.34	1.84	2.42	3.09	3.85	4.71	5.68	6.76	7.95	9.28	10.75	12.37	14.16	60
62	0.15	0.34	0.61	0.96	1.39	1.90	2.50	3.19	3.98	4.87	5.87	6.98	8.22	9.59	11.11	12.79	14.64	62
64	0.16	0.35	0.63	0.99	1.43	1.96	2.58	3.29	4.11	5.03	6.06	7.21	8.48	9.90	11.47	13.20	15.11	64
66	0.16	0.36	0.65	1.02	1.47	2.02	2.66	3.40	4.24	5.18	6.25	7.43	8.75	10.21	11.83	13.61	15.58	66
68	0.17	0.37	0.67	1.05	1.52	2.08	2.74	3.50	4.36	5.34	6.44	7.66	9.01	10.52	12.18	14.02	16.05	68
70	0.17	0.39	0.69	1.08	1.56	2.14	2.82	3.60	4.49	5.50	6.62	7.88	9.28	10.83	12.54	14.44	16.52	70
72	0.18	0.40	0.71	1.11	1.61	2.20	2.90	3.71	4.62	5.65	6.81	8.11	9.55	11.14	12.90	14.85	17.00	72
74	0.18	0.41	0.73	1.14	1.65	2.27	2.98	3.81	4.75	5.81	7.00	8.33	9.81	11.45	13.26	15.26	17.47	74
76	0.19	0.42	0.75	1.17	1.70	2.33	3.06	3.91	4.88	5.97	7.19	8.56	10.08	11.76	13.62	15.67	17.94	76
78	0.19	0.43	0.77	1.20	1.74	2.39	3.14	4.01	5.01	6.13	7.38	8.78	10.34	12.07	13.98	16.09	18.41	78
80	0.20	0.44	0.79	1.23	1.79	2.45	3.22	4.12	5.13	6.28	7.57	9.01	10.61	12.38	14.33	16.50	18.89	80
82	0.20	0.45	0.81	1.27	1.83	2.51	3.30	4.22	5.26	6.44	7.76	9.23	10.87	12.69	14.69	16.91	19.36	82
84	0.20	0.46	0.83	1.30	1.88	2.57	3.39	4.32	5.39	6.60	7.95	9.46	11.14	12.99	15.05	17.32	19.83	84
86	0.21	0.47	0.85	1.33	1.92	2.63	3.47	4.43	5.52	6.75	8.14	9.68	11.40	13.30	15.41	17.73	20.30	86
88	0.21	0.48	0.87	1.36	1.97	2.69	3.55	4.53	5.65	6.91	8.33	9.91	11.67	13.61	15.77	18.15	20.77	88
90	0.22	0.50	0.88	1.39	2.01	2.75	3.63	4.63	5.78	7.07	8.52	10.13	11.93	13.92	16.13	18.56	21.25	90
92	0.22	0.51	0.90	1.42	2.06	2.82	3.71	4.73	5.90	7.22	8.71	10.36	12.20	14.23	16.48	18.97	21.72	92
94	0.23	0.52	0.92	1.45	2.10	2.88	3.79	4.84	6.03	7.38	8.90	10.58	12.46	14.54	16.84	19.38	22.19	94
96	0.23	0.53	0.94	1.48	2.14	2.94	3.87	4.94	6.16	7.54	9.09	10.81	12.73	14.85	17.20	19.80	22.66	96
98	0.24	0.54	0.96	1.51	2.19	3.00	3.95	5.04	6.29	7.70	9.27	11.03	12.99	15.16	17.56	20.21	23.13	98
100	0.24	0.55	0.98	1.54	2.23	3.06	4.03	5.15	6.42	7.85	9.46	11.26	13.26	15.47	17.92	20.62	23.61	100

Table 3. Wet-line table (in feet) giving differences between wet-line length and vertical depth for selected vertical angles.

Wet-line length (feet)	4°	6°	8°	10°	12°	14°	16°	18°	20°	22°	24°	26°	28°	30°	32°	34°	36°	Wet-line length (feet)
10	0.01	0.02	0.03	0.05	0.07	0.10	0.13	0.16	0.20	0.25	0.30	0.35	0.41	0.47	0.54	0.62	0.70	10
12	0.01	0.02	0.04	0.06	0.09	0.12	0.15	0.20	0.24	0.30	0.36	0.42	0.49	0.57	0.65	0.74	0.84	12
14	0.01	0.02	0.04	0.07	0.10	0.14	0.18	0.23	0.29	0.35	0.41	0.49	0.57	0.66	0.76	0.87	0.98	14
16	0.01	0.03	0.05	0.08	0.12	0.16	0.20	0.26	0.33	0.40	0.47	0.56	0.65	0.76	0.87	0.99	1.12	16
18	0.01	0.03	0.06	0.09	0.13	0.18	0.23	0.30	0.37	0.45	0.53	0.63	0.73	0.85	0.98	1.12	1.26	18
20	0.01	0.03	0.06	0.10	0.14	0.20	0.26	0.33	0.41	0.50	0.59	0.70	0.82	0.94	1.09	1.24	1.40	20
22	0.01	0.04	0.07	0.11	0.16	0.22	0.28	0.36	0.45	0.55	0.65	0.77	0.90	1.04	1.20	1.36	1.54	22
24	0.01	0.04	0.08	0.12	0.17	0.24	0.31	0.39	0.49	0.60	0.71	0.84	0.98	1.13	1.31	1.49	1.68	24
26	0.02	0.04	0.08	0.13	0.19	0.25	0.33	0.43	0.53	0.64	0.77	0.91	1.06	1.23	1.41	1.61	1.81	26
28	0.02	0.04	0.09	0.14	0.20	0.27	0.36	0.46	0.57	0.69	0.83	0.98	1.14	1.32	1.52	1.74	1.95	28
30	0.02	0.05	0.10	0.15	0.22	0.29	0.38	0.49	0.61	0.74	0.89	1.05	1.22	1.42	1.63	1.86	2.09	30
32	0.02	0.05	0.10	0.16	0.23	0.31	0.41	0.52	0.65	0.79	0.95	1.12	1.31	1.51	1.74	1.98	2.23	32
34	0.02	0.05	0.11	0.17	0.24	0.33	0.44	0.56	0.69	0.84	1.01	1.19	1.39	1.60	1.85	2.11	2.37	34
36	0.02	0.06	0.12	0.18	0.26	0.35	0.46	0.59	0.73	0.89	1.07	1.26	1.47	1.70	1.96	2.23	2.51	36
38	0.02	0.06	0.12	0.19	0.27	0.37	0.49	0.62	0.78	0.94	1.12	1.33	1.55	1.79	2.07	2.36	2.65	38
40	0.02	0.06	0.13	0.20	0.29	0.39	0.51	0.66	0.82	0.99	1.18	1.40	1.63	1.89	2.18	2.48	2.79	40
42	0.03	0.07	0.13	0.21	0.30	0.41	0.54	0.69	0.86	1.04	1.24	1.47	1.71	1.98	2.28	2.60	2.93	42
44	0.03	0.07	0.14	0.22	0.32	0.43	0.56	0.72	0.90	1.09	1.30	1.54	1.80	2.08	2.39	2.73	3.07	44
46	0.03	0.07	0.15	0.23	0.33	0.45	0.59	0.75	0.94	1.14	1.36	1.61	1.88	2.17	2.50	2.85	3.21	46
48	0.03	0.08	0.15	0.24	0.35	0.47	0.61	0.79	0.98	1.19	1.42	1.68	1.96	2.27	2.61	2.98	3.35	48
50	0.03	0.08	0.16	0.25	0.36	0.49	0.64	0.82	1.02	1.24	1.48	1.75	2.04	2.36	2.72	3.10	3.49	50
52	0.03	0.08	0.17	0.26	0.37	0.51	0.67	0.85	1.06	1.29	1.54	1.82	2.12	2.45	2.83	3.22	3.63	52
54	0.03	0.09	0.17	0.27	0.39	0.53	0.69	0.89	1.10	1.34	1.60	1.89	2.20	2.55	2.94	3.35	3.77	54
56	0.03	0.09	0.18	0.28	0.40	0.55	0.72	0.92	1.14	1.39	1.66	1.96	2.28	2.64	3.05	3.47	3.91	56
58	0.03	0.09	0.19	0.29	0.42	0.57	0.74	0.95	1.18	1.44	1.72	2.03	2.37	2.74	3.16	3.60	4.05	58
60	0.04	0.10	0.19	0.30	0.43	0.59	0.77	0.98	1.22	1.49	1.78	2.10	2.45	2.83	3.26	3.72	4.19	60
62	0.04	0.10	0.20	0.31	0.45	0.61	0.79	1.02	1.26	1.54	1.84	2.17	2.53	2.93	3.37	3.84	4.33	62
64	0.04	0.10	0.20	0.32	0.46	0.63	0.82	1.05	1.31	1.59	1.89	2.24	2.61	3.02	3.48	3.97	4.47	64
66	0.04	0.11	0.21	0.33	0.48	0.65	0.84	1.08	1.35	1.64	1.95	2.31	2.69	3.12	3.59	4.09	4.61	66
68	0.04	0.11	0.22	0.34	0.49	0.67	0.87	1.12	1.39	1.69	2.01	2.38	2.77	3.21	3.70	4.22	4.75	68
70	0.04	0.11	0.22	0.35	0.50	0.69	0.90	1.15	1.43	1.74	2.07	2.45	2.86	3.30	3.81	4.34	4.89	70
72	0.04	0.12	0.23	0.36	0.52	0.71	0.92	1.18	1.47	1.79	2.13	2.52	2.94	3.40	3.92	4.46	5.03	72
74	0.04	0.12	0.24	0.37	0.53	0.73	0.95	1.21	1.51	1.84	2.19	2.59	3.02	3.49	4.03	4.59	5.17	74
76	0.05	0.12	0.24	0.38	0.55	0.74	0.97	1.25	1.55	1.88	2.25	2.66	3.10	3.59	4.13	4.71	5.30	76
78	0.05	0.12	0.25	0.39	0.56	0.76	1.00	1.28	1.59	1.93	2.31	2.73	3.18	3.68	4.24	4.84	5.44	78
80	0.05	0.13	0.25	0.40	0.58	0.78	1.02	1.31	1.63	1.98	2.37	2.80	3.26	3.78	4.35	4.96	5.58	80
82	0.05	0.13	0.26	0.41	0.59	0.80	1.05	1.34	1.67	2.03	2.43	2.87	3.35	3.87	4.46	5.08	5.72	82
84	0.05	0.13	0.27	0.42	0.60	0.82	1.08	1.38	1.71	2.08	2.49	2.94	3.43	3.96	4.57	5.21	5.86	84
86	0.05	0.14	0.28	0.43	0.62	0.84	1.10	1.41	1.75	2.13	2.55	3.01	3.51	4.06	4.68	5.33	6.00	86
88	0.05	0.14	0.28	0.44	0.63	0.86	1.13	1.44	1.80	2.18	2.60	3.08	3.59	4.15	4.79	5.46	6.14	88
90	0.05	0.14	0.29	0.45	0.65	0.88	1.15	1.48	1.84	2.23	2.66	3.15	3.67	4.25	4.90	5.58	6.28	90
92	0.06	0.15	0.29	0.46	0.66	0.90	1.18	1.51	1.88	2.28	2.72	3.22	3.75	4.34	5.00	5.70	6.42	92
94	0.06	0.15	0.30	0.47	0.68	0.92	1.20	1.54	1.92	2.33	2.78	3.29	3.84	4.44	5.11	5.83	6.56	94
96	0.06	0.15	0.31	0.48	0.69	0.94	1.23	1.57	1.96	2.38	2.84	3.36	3.92	4.53	5.22	5.95	6.70	96
98	0.06	0.16	0.31	0.49	0.71	0.96	1.25	1.61	2.00	2.43	2.90	3.43	4.00	4.63	5.33	6.08	6.84	98
100	0.06	0.16	0.32	0.50	0.72	0.98	1.28	1.64	2.04	2.48	2.96	3.50	4.08	4.72	5.44	6.20	6.98	100

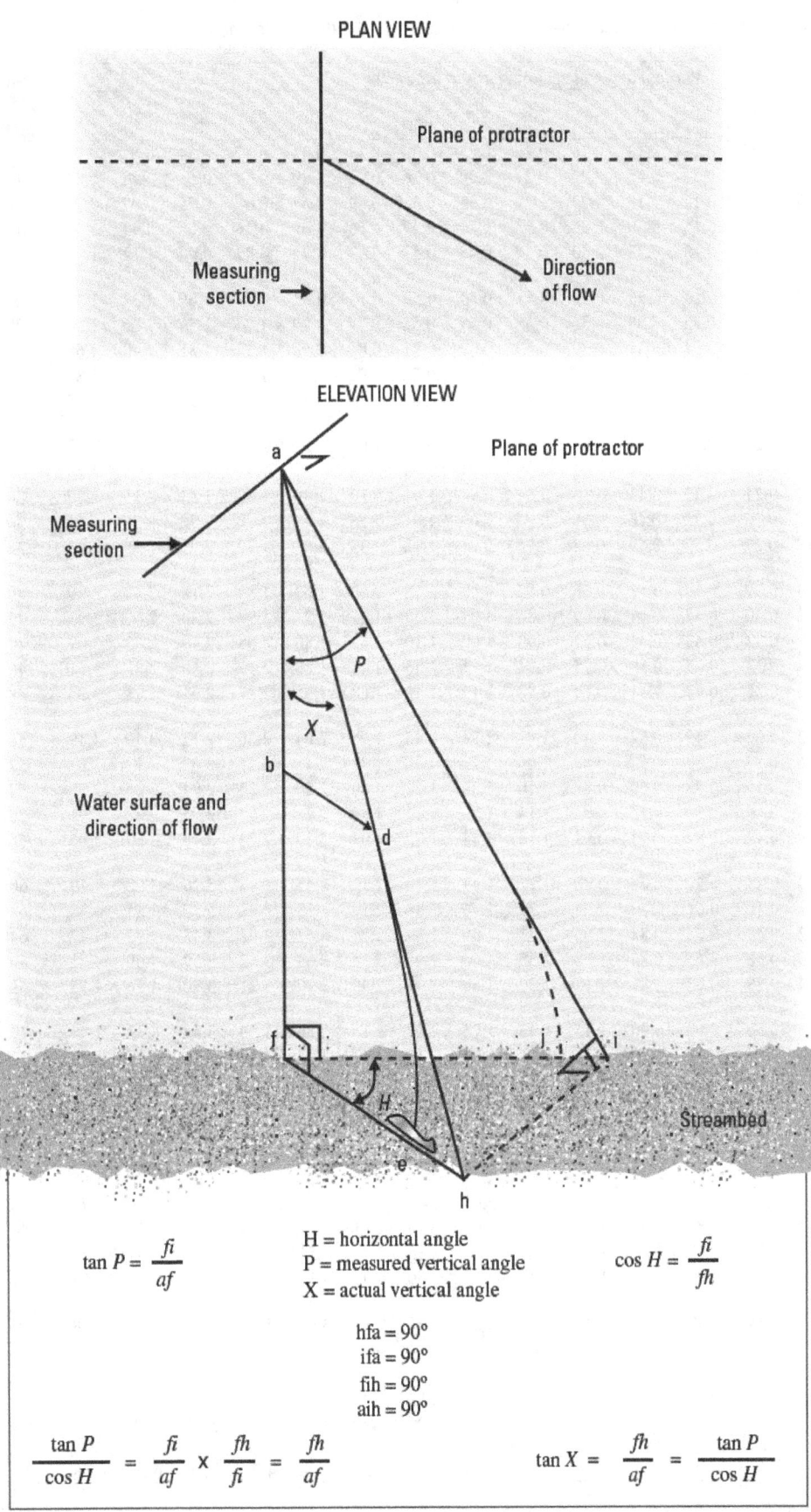

PLAN VIEW

Plane of protractor

Measuring section →

Direction of flow

ELEVATION VIEW

a

Plane of protractor

Measuring section →

P

X

b

Water surface and direction of flow

d

f

j   i

H

e

Streambed

h

$$\tan P = \frac{fi}{af}$$

H = horizontal angle
P = measured vertical angle
X = actual vertical angle

$$\cos H = \frac{fi}{fh}$$

hfa = 90°
ifa = 90°
fih = 90°
aih = 90°

$$\frac{\tan P}{\cos H} = \frac{fi}{af} \times \frac{fh}{fi} = \frac{fh}{af}$$

$$\tan X = \frac{fh}{af} = \frac{\tan P}{\cos H}$$

Figure 11. Sketch of geometric relationship of actual to measured vertical angle when flow direction is not normal to the measuring section.

5. If the angle increases appreciably when the additional line is let out, let out more line until the total additional line, the angle, and the vertical distance are in agreement with values presented in the air-correction table.

To place the meter at the 0.8-depth position, a correction to the amount of line reeled in must be made for the difference, if any, between the air correction for the sounding position and that for the 0.8-depth position. This difference is designated as *m* in table 5. If the angle increases for the 0.8-depth position, the meter must be lowered; if it decreases, the meter must be raised.

For the 0.8-depth position of the meter, the wet-line correction may require consideration if the depths are more than 40 ft and if the change in vertical angle is more than 5 percent. If the vertical angle remains the same or decreases, the wet-line correction (table 3) for the 0.8-depth position is less than the wet-line correction for the sounding position by some difference, designated as *n* in table 5. If the vertical angle increases, the difference in correction *n* diminishes until the increase in angle is about 10 percent; for greater increases in angle, the difference between corrections increases also. Table 5 summarizes the effect on air and wet-line corrections caused by raising the meter from the sounding position to the 0.8-depth position.

For slight changes in the vertical angle, because of the differences *m* and *n* in the air and wet-line corrections, the adjustments

to the wet-line length of the 0.8-depth position are small and usually can be ignored. Table 5 indicates that the meter may be placed a little too deep if the adjustments are not made. Because of this possibility, the wet-line depth instead of the vertical depth is sometimes used as the basis for computing the 0.8-depth position with no adjustments for the differences *m* and *n*.

Use of Sonic Sounder

The sonic sounder has been used primarily for measuring depth when making a moving boat measurement and in ADCP discharge measurements, and is generally not utilized for measurements where sounding weights are used. However, it can be used in swift, debris-laden streams, where it is difficult or dangerous to lower the sounding weight and meter into the water. The sonic sounder will record the depth when the weight is just below the water surface. For moving boat measurements, the sonic sounder records a continuous trace of the streambed on a digital or analog chart. Details of the setup and use can be found in Smoot and Novak (1969). Relevant information on the use of acoustic sounders in a riverine environment is discussed in detail in Mueller and Landers (1999) and by the International Organization for Standardization (2003). For use of a sonic sounder in ADCP measurements, consult Mueller and Wagner (2009).

Table 4. Degrees to be added to observed vertical angle, *P*, to obtain true vertical angle when flow direction is not normal to measurement section.

Observed vertical angle, *P*, in degrees	Horizontal angle, *H*, in degrees					
	8 cos = 0.99	12 cos = 0.98	16 cos = 0.96	20 cos = 0.94	24 cos = 0.91	28 cos = 0.88
8	0.1	0.2	0.3	0.5	0.8	1.0
12	0.1	0.3	0.5	0.8	1.1	1.5
16	0.1	0.4	0.6	1.0	1.4	2.0
20	0.2	0.4	0.7	1.2	1.7	2.4
24	0.2	0.5	0.8	1.4	2.0	2.8
28	0.2	0.5	1.0	1.5	2.2	3.0
32	0.2	0.6	1.0	1.6	2.4	3.3
36	0.2	0.6	1.1	1.7	2.5	3.4

Table 5. Summary table for setting the meter at 0.8-depth position in deep, swift streams.

Change in vertical angle	Air correction		Wet-line correction	
	Direction of change	Correction to meter position	Direction of change	Correction to meter position
None	None	None	Decrease	Raise meter the distance *n*
Decrease	Decrease	Raise meter the distance *m*	Decrease	Raise meter the distance *n*
Increase	Increase	Lower meter the distance *m*	Decrease, then increase	[1]

[1]Raise meter the distance *n* unless the increase in angle is greater than about 10 percent, then it is necessary to lower the meter the distance *n*.

Measurement of Velocity

With point-velocity meters, after the width and depth at a vertical are measured and recorded, determine the method of velocity measurement. Normally the two-point method or the 0.6-depth method is used. Details of velocity measurement methods using various equipment and under different flow conditions are described in subsequent sections of this chapter.

Compute the setting of the meter for the particular method that will be used at that depth. For the top-setting wading rod, or the spiral-computing dial, some meter settings are self-computing. Record the meter position as 0.8, 0.6, 0.2, or another setting. After the meter is placed at the proper depth, let it adjust to the current before starting the velocity observation. With an ADV, make sure the instrument is located perpendicular to the tag line and the wading rod is plumbed (that is, vertical to the channel bed) before beginning measurement. The time required for adjustment to the undisturbed stream velocity is usually only a few seconds if the velocities are greater than 1 ft/s; however, for lower velocities, particularly if the current meter is suspended by a cable, a longer period of adjustment is needed. After the meter has become adjusted to the current, count the number of revolutions made by the rotor for a period of 40 to 70 seconds; for an ADV, this time period is typically programmed into the instrument's discharge measurement software.

If using a stopwatch to time the revolutions of a mechanical current meter, start the stopwatch simultaneously with the first signal or click, counting "zero," not "one." End the count on a convenient number given in the mechanical current meter rating table column heading. Stop the stopwatch on that count and read the time to the nearest second. Record the number of revolutions and the time interval. If the velocity is to be observed at more than one point in the vertical, determine the meter setting for the additional observation, set the meter to that depth, time the revolutions, and record the data.

When using a current meter digitizer (CMD), a personal digital assistant (PDA), or an electronic notebook such as the Aquacalc or an ADV, observe the same basic procedure for setting the meter, and for providing time for the meter to stabilize. With these instruments, however, the counting and timing of the rotor revolutions or the acoustic pulse measurement are performed automatically. The number of revolutions, time, and velocity displayed by the CMD must be transferred manually to paper field notes, whereas, these data are electronically recorded by the Aquacalc, a PDA, or an ADV. When using any of these automatic meter counting devices and a mechanical current meter, make sure that multiple counts are not occurring during measurement of slow velocity. This can sometimes be determined by visually observing the rotation of the rotor while simultaneously listening to the audible clicks or beeps from the counting device. With the ADV, instead of audible clicks, acoustic Doppler theory is applied. The ADV can sense and measure velocities much smaller than those

rated and measured by mechanical meters and are not prone to the multiple count errors of the mechanical current meter.

Current meters, in general, measure stream velocity at a point. One notable exception is the ADCP. This method will be discussed in a subsequent section of this chapter; a thorough discussion of a moving boat discharge measurement using an ADCP is documented in Mueller and Wagner (2009).

The method of making discharge measurements at a cross section by using a current meter that measures point velocities requires determination of the mean velocity in each of the selected verticals. The mean velocity in a vertical is obtained from velocity observations at several points in that vertical. The mean velocity can be approximated by making a few velocity observations and using a known relation between those velocities and the mean in the vertical. The various methods of measuring velocity are: vertical-velocity curve, two-point, 0.6-depth, 0.2-depth, three-point, and surface and subsurface.

Vertical-Velocity Curve Method

In the vertical-velocity curve method, a series of velocity observations at points well distributed between the water surface and the streambed are made at each of the verticals. If there is considerable curvature in the lower part of the vertical-velocity curve, then it is advisable to space the observations more closely together in that part of the depth. Normally, the observations are taken at 0.1-depth increments between 0.1 and 0.9 of the depth. Observations are always taken at 0.2, 0.6, and 0.8 of the depth so that the results obtained by the vertical-velocity curve method may be compared with the commonly used methods of velocity observation. Observations are made at least 0.5 ft from the water surface and from the streambed with the Price AA meter, at least 0.3 ft from these boundaries with the Price pygmy meter, or at least 0.2 ft from these boundaries with the FlowTracker ADV.

The vertical-velocity curve for each vertical is based on observed velocities plotted against depth, as shown in figure 12. In order that vertical-velocity curves at different verticals may be readily compared, it is customary to plot depths as proportional parts of the total depth. The mean velocity in the vertical is obtained by measuring the area between the curve and the ordinate axis with a planimeter, or by other means, and dividing the area by the length of the ordinate axis.

The vertical-velocity curve method is valuable in determining coefficients for application to the results obtained by other methods, but is not generally adapted to routine discharge measurements because of the extra time required to collect field data and to compute the mean velocity. A typical vertical-velocity curve for the cross section should be measured and evaluated at all new measurement sites, and perhaps at new measurement sections if the new section is significantly different in hydraulic characteristics than the section normally used at a regular measurement site.

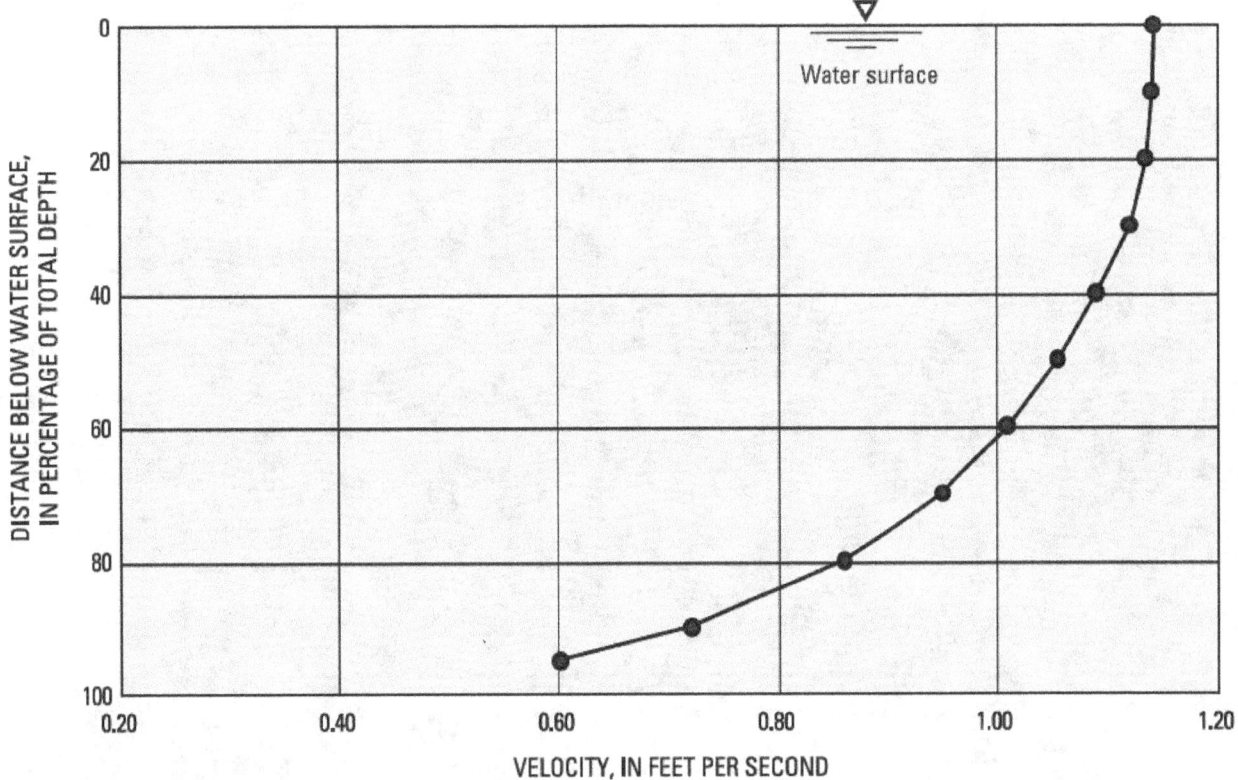

Figure 12.   Typical vertical-velocity curve.

Two-Point Method

This is the preferred method for making midsection discharge measurements with point velocity meters. In the two-point method of measuring velocities, observations are made in each vertical at 0.2 and 0.8 of the depth below the surface. The average of these two observations is used as the mean velocity in the vertical. This method is based on many studies of actual observation and on mathematical theory. Experience has shown that this method gives more consistent and accurate results than any of the other methods, except for the vertical-velocity curve method. Use the two-point method for depths of 2.5 ft or greater, unless using a pygmy current meter or an ADV, in which case, this method is used in depths of 1.5 ft or greater.

With an ADV, the actual instrument has much less drag or resistance to the flow as compared to a mechanical current meter. To prevent boundary interference, avoid placing the ADV sample volume [typically 10 centimeters (about 4 in.) from the center transmitting transducer] within 2 in. from any solid boundary. This boundary condition with the ADV allows for measurement of velocity closer to the water surface and channel bed than a Price AA or to the pygmy (fig. 13). With the Price AA current meter, the two-point method is not used at depths less than 2.5 ft because the current meter would be too close to the water surface and to the streambed to give dependable results.

Six-Tenths-Depth Method

In the 0.6-depth method, an observation of velocity made at 0.6 of the depth below the water surface in the vertical is used as the mean velocity in the vertical. Actual observation and mathematical theory have shown that the 0.6-depth method gives reliable results and is used by the USGS under the following conditions:

Price AA Current Meter

1. Whenever the depth is between 0.3 ft and 2.5 ft.

2. When large amounts of slush ice or debris make it impossible to observe the velocity accurately at the 0.2 depth. (This condition prevents the use of the two-point method.)

3. When the meter is placed a distance above the sounding weight, which makes it impossible to place the meter at the 0.8 depth. (This condition prevents the use of the two-point method.)

4. When the stage in a stream is changing rapidly and you must make a quick measurement.

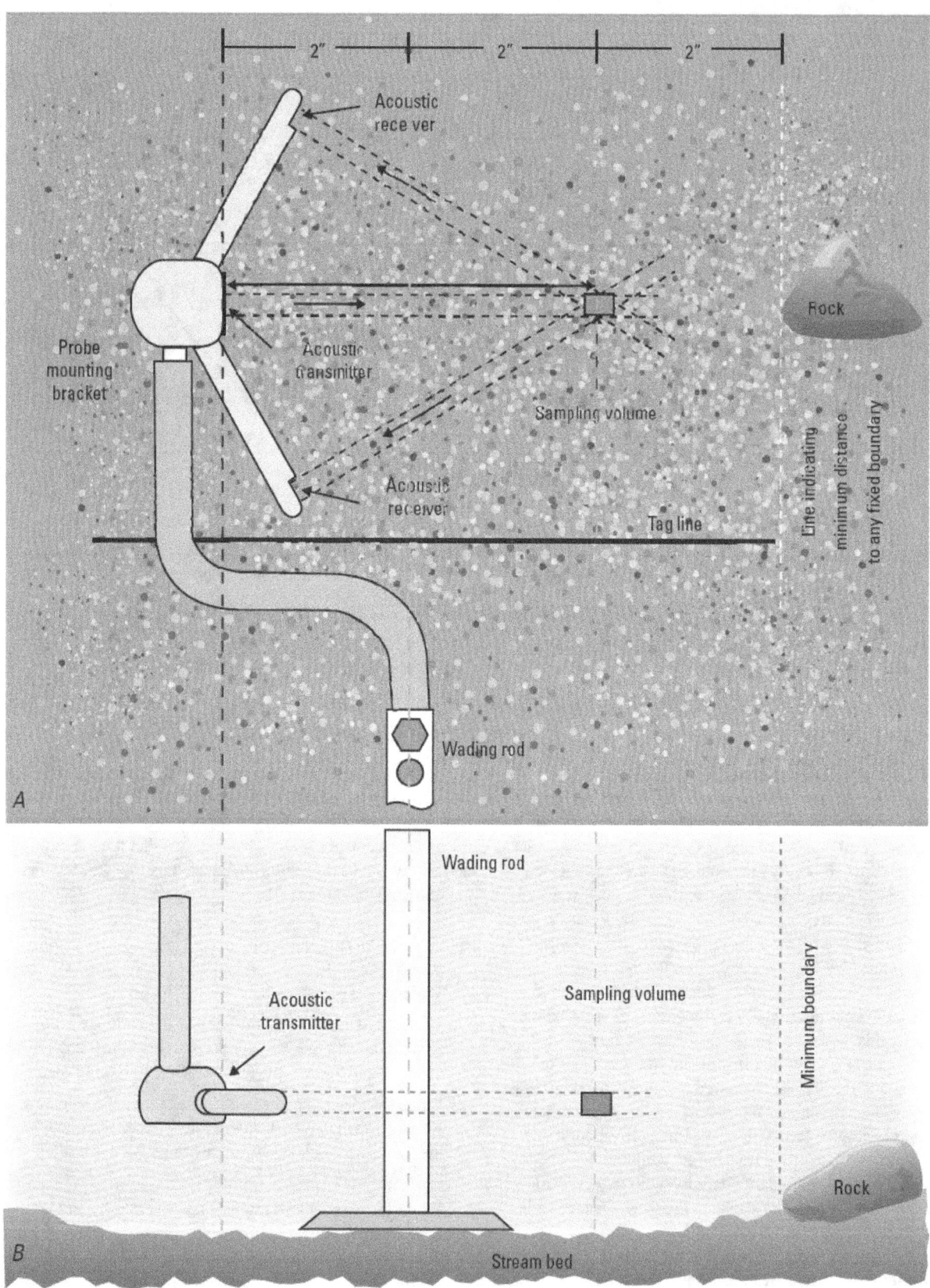

Figure 13.   Schematic showing A, plan vie and B, cross section of the FlowTracker sampling volume and the proximity to a fixed boundary in the stream. (Note: Probe mounting bracket not shown.)

Pygmy Current Meter

1. Whenever the depth is between 0.3 ft and 1.5 ft.
2. When large amounts of slush ice or debris make it impossible to observe the velocity accurately at the 0.2 depth. (This condition prevents the use of the two-point method.)
3. When the stage in a stream is changing rapidly and a measurement must be made quickly.

Acoustic Doppler Velocimeter

1. Whenever the depth is between 0.25 ft and 1.5 ft.
2. When large amounts of slush ice or debris make it impossible to observe the velocity accurately at the 0.2 depth. (This condition prevents the use of the two-point method.)
3. When the stage in a stream is changing rapidly and a measurement must be made quickly.

Two-Tenths-Depth Method

The 0.2-depth method consists of observing the velocity at 0.2 of the depth below the surface and applying a coefficient to this observed velocity to obtain the mean in the vertical. It is used mainly during times of high water when the velocities are great, making it impossible to obtain soundings or to place the meter at the 0.8 or the 0.6 depth.

Use a standard cross section or a general knowledge of the cross section at a site to compute the 0.2 depth when it is impossible to obtain depth soundings. A sizeable error in an assumed 0.2 depth is not critical because the slope of the vertical-velocity curve at this point is usually nearly vertical. The 0.2 depth is also used in conjunction with the sonic sounder for flood measurements. The two-point method and the 0.6-depth method are preferred over the 0.2-depth method because of their greater accuracy.

The discharge measurement is normally computed by using the 0.2-depth velocity observations without coefficients as though each were a mean in the vertical. The approximate discharge thus obtained divided by the area of the measuring section gives the weighted mean value of the 0.2-depth velocity. Studies of many measurements made by the two-point method show that for a given measuring section, the relation between the mean 0.2-depth velocity and the true mean velocity either remains constant or varies uniformly with stage. In either circumstance, this relation may be determined for a particular 0.2-depth measurement by recomputing measurements made at the site by the two-point method using only the 0.2-depth velocity observation as the mean in the vertical. The plotting of the true mean velocity versus the mean 0.2-depth velocity for each measurement will give a velocity-relation curve for use in adjusting the mean velocity for measurements made by the 0.2-depth method.

If not enough measurements by the two-point method are available at a site to establish a velocity-relation curve, vertical-velocity curves are needed to establish a relation between the mean velocity and the 0.2-depth velocity. The usual coefficient to adjust the 0.2-depth velocity to the mean velocity is about 0.88.

Three-Point Method

The three-point method consists of observing the velocity at 0.2, 0.6, and 0.8 of the depth, thereby combining the two-point and 0.6-depth methods. The preferred method of computing the mean velocity is to average the 0.2- and 0.8-depth observations and then average this result with the 0.6-depth observation. However, when more weight to the 0.2- and 0.8-depth observations is desired, the arithmetic mean of the three observations may be used.

The three-point method is used when the velocities in the vertical are abnormally distributed [for example, the 0.2 (top) velocity is more than twice the 0.8 (bottom) velocity, or the 0.8 (bottom) velocity is greater than the 0.2 (top) velocity]. It is also used when the 0.8-depth observation is made where the velocity is seriously affected by friction or by turbulence produced by the streambed or an obstruction in the stream. If using a Price AA, the depths must be greater than 2.5 ft to use this method. If using a Price pygmy or ADV, the depths must be greater than 1.5 ft to use this method.

Surface and Subsurface Methods

Surface and subsurface methods consist of observing the velocity at the water surface or some distance below the water surface. Surface measurements may be made with the optical current meter, or by observing and timing surface floats. Subsurface measurements are made with a current meter at a distance of at least 2 ft below the surface to avoid the effect of surface disturbances. Surface and subsurface measurements are used primarily for deep swift streams where it is impossible or dangerous to obtain depth and velocity soundings at the regular 0.2, 0.6, and 0.8 depths.

Coefficients are necessary to convert the surface or subsurface velocities to the mean velocity in the vertical. Vertical-velocity curves obtained at the particular site are the best method to compute these coefficients. However, the coefficients are generally difficult to determine reliably because they may vary with stage, depth, and position in the measuring cross section. Experience has shown that the coefficients generally range from about 0.84 to about 0.90, depending on the shape of the vertical-velocity curve. The higher values are usually associated with smooth streambeds and normally shaped vertical-velocity curves, whereas the lower values are associated with irregular streambeds and irregular vertical-velocity curves.

Direction of Flow Measurements

Consider the direction of flow because the component of velocity normal to the measurement section is that which must be determined by both mechanical and acoustic Doppler point-velocity current meters. Generally, for the mechanical meter, the relation for velocity components not normal to the measuring section can be visualized in figure 14, and should be corrected using the cosine of alpha.

Flow direction is also critical with respect to an ADV, since the ADV assumes a horizontal and perpendicular plane to the flow. The hydrographer must pay close attention to the flow angle reported by the FlowTracker. Always hold the wading rod (with FlowTracker attached) perpendicular to the tag line so that the pulse generated by the transmitter is parallel to the tag line. Ideally, the tag line should be set up in the cross section to be measured so that flow is perpendicular to the tag line. Flow angle, as calculated by the FlowTracker, is defined as the direction of flow relative to the *x*-direction of flow, so that:

$$FlowAngle = \arctan(Vy / Vx), \qquad (10)$$

where Vy is the velocity in the *y* direction (parallel to the tag line), and

Vx is the velocity in the *x* direction (perpendicular to the tag line) used to calculate discharge.

The flow angle calculated by the FlowTracker can result from two sources: (1) the flow is not perpendicular to the tag line, and (2) the flow is perpendicular to the tag line but the wading rod is not held correctly relative to the tag line, as described above. Regarding source (1), some small angles and variation in the flow angle at a site is not unusual. However, if large fluctuations of flow angles are reported, make measurements at another section with more uniform flow. Regarding source (2), holding the FlowTracker so that it is skewed at any angle relative to the tag line will result in a measurement of velocity that is biased low. Small angles do not result in significant biases, but because of these biases, users should be careful to minimize this error. If the FlowTracker is held so that it is skewed at an angle of approximately 8 degrees from the tag line, the measured velocity may be in error by as much as 1 percent (assuming that flow is perpendicular to the tag line). Large variations in flow angles may be indicative of poor or inconsistent alignment of the wading rod or poor site selection for the measurement.

In a wading measurement, if the meter used is a horizontal-axis meter with a component propeller, such as the Ott meter, the propeller should be pointed upstream at right angles to the cross section, but only if alpha is less than 45 degrees. Such a meter will register the desired component of velocity normal to the cross section when alpha is less than 45 degrees. The same procedure should be used if an electromagnetic component meter is used. These meters also measure the component of velocity normal to the measuring section. Generally, for either type of meter, if alpha is greater than 45 degrees, the component meter should be pointed directly into the current,

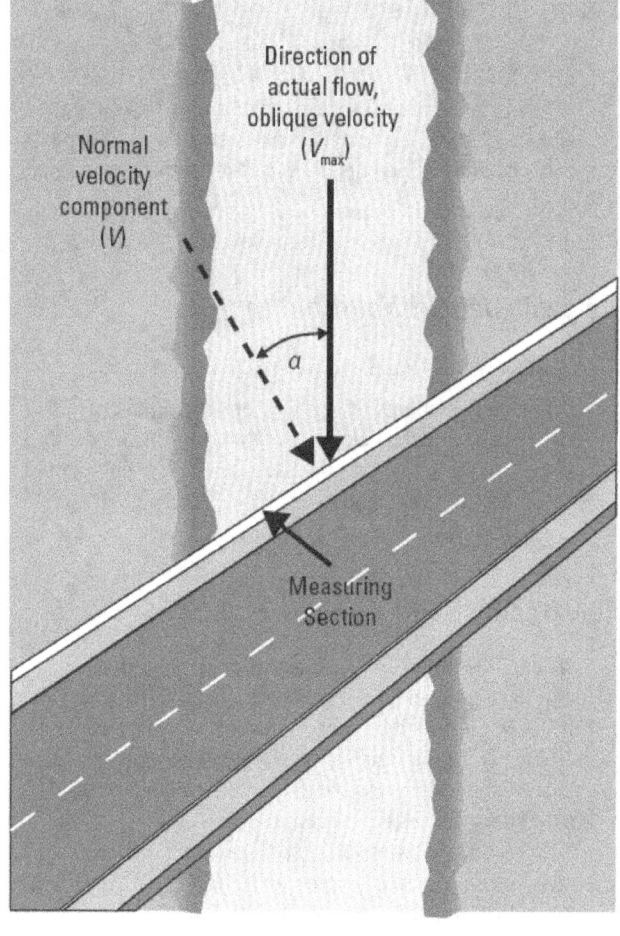

$$V = V_{max} \times \cos a$$

Figure 14. Velocity components when flow is not normal to measuring section.

and the horizontal angle correction should be applied as described in the following paragraphs.

Other meters on a wading-rod suspension, such as the vertical-axis Price current meter, should be pointed into the current. Any meter on a cable suspension will automatically point into the current because of the effect of the meter vanes. When the meter is pointed into an oblique current, the measured velocity must be multiplied by the cosine of the angle (alpha) between the current and a perpendicular to the measurement section in order to obtain the desired normal component of the velocity.

Either of two methods may be used to obtain the cosine of the angle alpha. In the first method, use the field note sheet that has a point of origin (*o*) printed on the left margin and cosine values on the right margin (see figure 24). Measure the cosine of the angle of the current by holding the note sheet in a horizontal position with the point of origin on the tag line, bridge rail, cable rail, or any other feature parallel to the cross section, as shown in figure 15. With the long side of the note sheet parallel to the direction of flow, the tag line or bridge rail

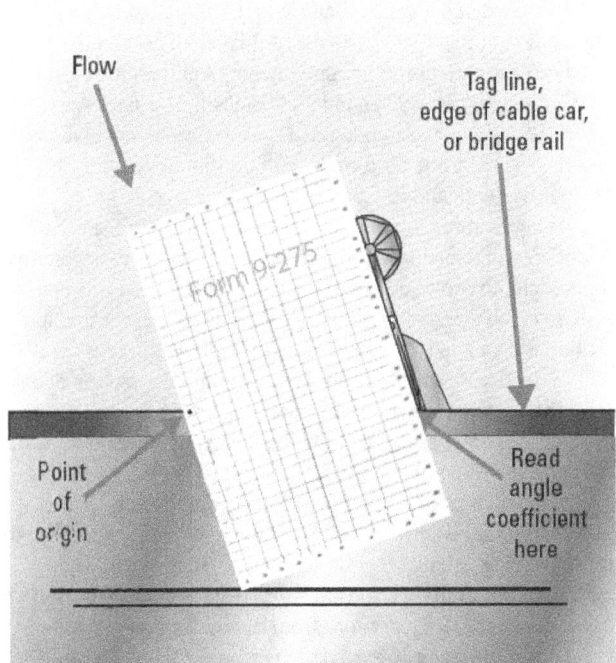

Figure 15. Measurement of horizontal angle with measurement note sheet.

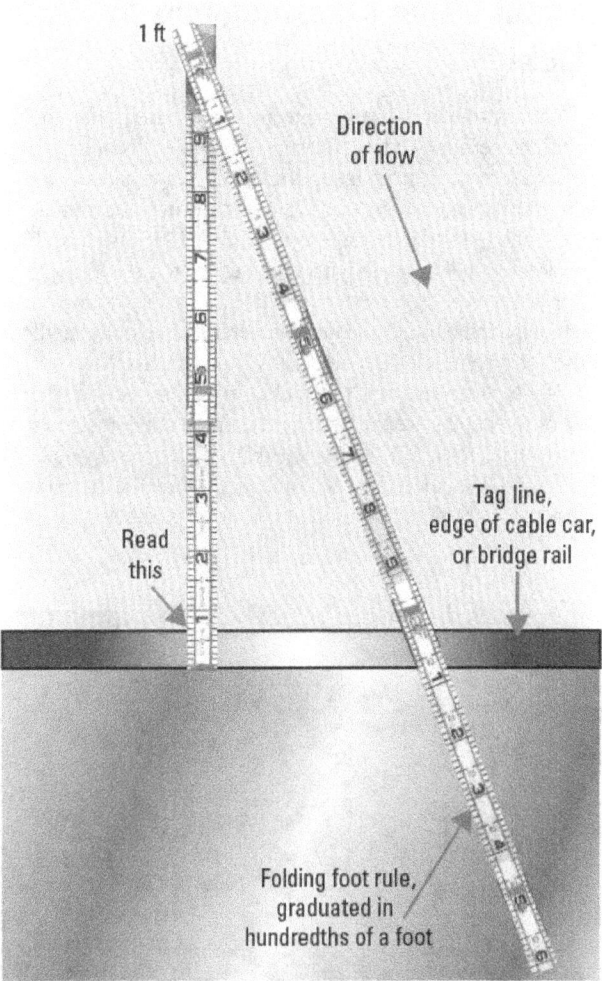

Figure 16. Measurement of horizontal angle with a folding rule.

will intersect the value of cosine alpha on the top, bottom, or right edge of the note sheet. The direction of the current will be apparent from the direction of movement of floating particles. If the water is clear of floating material, the edge of the note sheet is aligned parallel to the direction of movement. If no such material is available, the rather inelegant, but time-honored method of spitting into the stream can be used to discern the direction of flow. The position of the current meter may also be used if it can be seen below the water surface. Multiply the measured velocity by the cosine of the angle to determine the velocity component normal to the measurement section.

The second method of obtaining the cosine of the angle alpha involves the use of a folding rule that folds at 0.5-ft or 1-ft intervals. The rule must be graduated in hundredths of a foot and jointed every 0.5 ft or 1 ft. Extend the first 2 ft of the rule and place the 2.00-ft marker on the tag line or bridge rail, as shown in figure 16, with the rule aligned with the direction of flow. Fold the rule at the 1-ft mark so that the first foot of the rule is normal to the tag line or bridge rail. Make a reading where the 1-ft section intersects the tag line or bridge rail. That reading, subtracted from 1.00, is the cosine of the angle alpha. For example, if the reading on the rule is 0.07 ft (fig. 16), the cosine of alpha equals 0.93.

The direction of flow, as observed on the surface of the stream, may not always be a reliable indication of the direction of flow at some distance below the surface. For instance, when measuring a stream influenced by tidal fluctuations, it is possible to have flow moving downstream near the surface of the stream, and flow moving upstream near the bottom of the

stream. Therefore, whenever it is suspected that the direction of flow is variable at different depths, other means than those already mentioned must be used to determine the direction of flow. One such method is to use a rigid rod or pole, with a vane attached to the bottom, and an indicator parallel to the vane attached to the top. Another method is to use a sounding weight with a compass and remote readout, as described in the equipment section of this chapter. If the variation of the direction of flow in the vertical is not great, then an average value of the cosine of the angle may be used for computing the component mean velocity for the vertical. However, if the variation is considerable, then you may need to subdivide the vertical and make separate computations for each subdivision. This may require additional measurements of velocity in the vertical.

If available, an ADCP from a moving boat can be used to quickly discern multidirectional flow in the vertical and across the selected cross-section. In a tidal affected reach, an ADCP is the preferred method of measuring discharge because of its use in the measurement of three dimensional velocity through most of the water column.

Current-Meter Measurements by Wading

Current-meter measurements by wading are preferred, if conditions permit. Wading measurements offer the advantage over measurements from bridges and cableways because the hydrographer can usually chose the best of several available cross sections for the measurement. Figure 17 shows a wading measurement being made with a top-setting rod.

Use the type AA, pygmy, or ADV meter for wading measurements. Table 6 lists the type of meter and velocity method to use for wading measurements at various depths.

If a type AA meter is being used in a cross section where most of the depths are greater than 1.5 ft, do not change to the pygmy meter for a few depths less than 1.5 ft or vice versa. The Price AA meter is not recommended for depths of 1.0 ft or less because the registration of the meter is affected by its proximity to the water surface and to the streambed. However, it can be used at depths as shallow as 0.5 ft to avoid changing meters if only a few verticals of this depth are required. The type AA meter or the pygmy meter should not be used in velocities less than 0.2 ft/s unless it is absolutely necessary.

It is no longer recommended to use coefficients given by Pierce (1941) for the performance of current meters in water of shallow depth and low velocities.

When natural conditions for measuring are in the range considered undependable, modify the measuring cross section, if practical, to provide acceptable conditions. Often it is possible in small streams to build dikes to cut off dead water and shallow flows in a cross section, or to improve the cross section by removing the rocks and debris within the section and from the reach of stream immediately upstream from it. After modifying a cross section, allow the flow to stabilize before starting the discharge measurement.

Stand in a position that least affects the velocity of the water passing the current meter by facing the bank, with the water flowing against the side of the leg. Holding the wading rod at the tag line, stand from 1 to 3 in. downstream from the tag line and 18 in. or more from the wading rod. Avoid standing in the water if feet and legs would occupy a considerable percentage of the cross section of a narrow stream. In small streams where the width permits, stand on a plank or other support above the water rather than in the water. Velocity bias caused by effects of the hydrographer's position can be significant. Observance of these conditions is important while using mechanical meters, ADVs, and any wading measurement where an obstacle could interfere with the natural flow conditions of the stream.

When using a Price meter, keep the wading rod in a vertical position and the meter parallel to the direction of flow while observing the velocity. If the flow is not at right angles to the tag line, measure the angle coefficient carefully. When using an ADV or other instrument that can measure the x component velocity, the instrument should be aligned more precisely with the tag line. See the discussion of FlowTracker use and flow angles in the "Measurement of Velocity" section of this chapter.

During measurements of streams with shifting beds, the scoured depressions left by the hydrographer's feet can affect soundings or velocities. Generally, place the meter ahead of and upstream from the hydrographer's body and feet. Record an accurate description of streambed and water-surface configuration each time a discharge measurement is made in a sand-channel stream.

For discharge measurements of flow too small to measure with a current meter, use a volumetric method, Parshall flume, or weir plate. Those methods are described in subsequent sections of this chapter.

Figure 17. Wading measurement using a top-setting rod.

Table 6. Current meter and velocity-measurement method for various depths.

Depth, in feet	Current meter	Velocity method
2.5 and greater	Price Type AA	0.2 and 0.8
1.5 - 2.5	Price Type AA	0.6
0.3 - 1.5	Price Pygmy	0.6
1.5 and greater	Price Pygmy	0.2 and 0.8
0.3 - 1.5	ADV	0.6
1.5 and greater	ADV	0.2 and 0.8

Current-Meter Measurements From Cableways

The Price type-AA current meter is generally used in conjunction with sounding weights and a sounding reel when measuring discharge from a cableway, although in recent years, using an ADCP mounted to a tethered craft has become much more widespread. Stationing (for width measurements) is usually determined from marks painted on the cableway. The velocity is measured by setting the meter at the proper position in the vertical, as indicated in table 7. Table 7 is designed so that no velocity observations will be made with the meter closer than 0.5 ft to the water surface. In the zone from the water surface to a depth of 0.5 ft, the current meter is known to give erratic results.

One problem found while measuring velocities from a cableway is that the movement of the cable car from one station to the next causes the car to oscillate for a short time after coming to a stop. Wait until this oscillation has decreased to a negligible amount before counting the revolutions.

By using a method of tagging the sounding cable at convenient intervals with streamers of different-colored binding tapes, each colored streamer being a known distance above the current-meter rotor (known as using tags), the meter can be kept under water at all times to prevent it from freezing in cold air. Tags are also used in measurements of deep, swift streams. See the section of this chapter on "Measurement of depth."

If large amounts of debris are flowing in the stream, raise the meter up to the cable car several times during the measurement to be certain the pivot and rotor of the meter are free of debris. However, keep the meter in the water during the measurement if the air temperature is considerably below freezing.

During floods, there is always a danger of catching a submerged or floating object, such as a tree or log, which can endanger the sounding equipment, meter, and most importantly, the hydrographer. Always be sure that the sounding cable has been installed on the sounding reel, according to the breaking loads specified in table 8. This assures that the sounding cable will break when it reaches its end, thereby preventing a potentially serious accident where the cable car and hydrographer could be spilled into the stream. Also, for added safety, always carry a pair of lineman's side-cutter pliers while making measurements from a cableway. If the sounding cable becomes hopelessly hung and does not break, as it should, cut the sounding line to ensure safety. Sometimes the cable car can be pulled to the edge of the water and the debris can be released.

When measurements are made from cableways where the stream is deep and swift, measure the angle that the meter suspension cable makes with the vertical due to the drag. The vertical angle, measured by protractor, is needed to correct the soundings to obtain the actual vertical depth, as described in the section on "Depth corrections for downstream drift of current meter and weight."

Table 7. Velocity-measurement method for various suspensions and depths.

Suspension	Minimum depth, in feet	
	0.6 method	0.2 and 0.8 method
15 C .5[1], 30 C .5	1.2	2.5
50 C .55	1.4	2.8
50 C .9	2.2	4.5
75 C 1.0, 100 C 1.0, 150 C 1.0	2.5	5.0
200 C 1.5, 300 C 1.5	3.8[2]	7.5

[1]15 pound Columbus-type weight, 0.5 above channel bed.

[2]Use 0.2 method for depths 2.5 to 3.7 feet with appropriate coefficient (for example, 0.87 or 0.88).

Table 8. Breaking loads for Ellsworth stranded cable.

Sounding cable	Diameter, in inches	Total number of strands	Rated total breaking load, in pounds	Recommended breaking load, in pounds	Number of strands to cut	Number of strands to remain
Ellsworth 0.084	0.084	36	500	250	15	21
Ellsworth 0.100	0.100	30	1,000	500	15	15
Ellsworth 0.125	0.125	30	1,500	500	20	10

Current-Meter Measurements From Bridges

When a stream cannot be waded, a bridge may be used to obtain current-meter measurements. Many measuring sections under bridges are satisfactory for current-meter measurements, but cableway sections are usually better because they provide an unobstructed reach of the channel. In addition, cableways usually have no bridges constricting the free flow of the stream in the measuring reach.

No set rule can be given for choosing between the upstream or downstream side of the bridge while making a discharge measurement. The advantages of using the upstream side of the bridge are the following:

- Hydraulic characteristics at the upstream side of bridge openings usually are more favorable. Flow is usually smoother and there is less turbulence than at the downstream side of the bridge.

- Approaching drift can be seen and be more easily avoided.

- The streambed at the upstream side of the bridge is not likely to scour as much as at the downstream side.

The advantages of using the downstream side of the bridge are:

- Vertical angles are more easily measured because the sounding line will move away from the bridge.

- The flow lines of the stream may be straightened out by passing through a bridge opening with piers.

Using the upstream side or the downstream side of a bridge for a current-meter or ADCP measurement should be decided based on circumstances for each bridge. Consider the factors mentioned above and the physical conditions at the bridge, such as location of the walkway, traffic hazards, and accumulation of trash on piles and piers.

For an ADCP measurement with a tethered craft, unless a special rigid support for deployment or bank-operated cableway has been developed for the upstream side of the bridge, the downstream side of the bridge is usually where the ADCP is most conveniently deployed. Bridge piers can cause excessive turbulence during high streamflow, especially if debris accumulates on the piers and(or) the piers are skewed to the flow. The effect of bridge-pier-induced turbulence may be reduced when deploying an ADCP from the downstream side of the bridge by lengthening the tether to increase the distance between the bridge and the tethered boat. Close attention should be paid to the cross section to ensure that no large eddies that could cause flow to be nonhomogeneous are present. Possible alternatives to measuring off the downstream side of the bridges include using a bank-operated cableway, a rigid extension that allows the hydrographer to deploy from the upstream side of the bridge, or having personnel on each bank hold a rope or cord attached to the platform to pull the tethered boat back and forth across the river.

For a mechanical current meter measurement, use either a handline, or a sounding reel supported by a bridge board or a portable crane, to suspend the current meter and sounding weight from bridges. Depth measurements should be made as described in the section entitled "Measurement of depth." Measure the velocity by setting the meter at the position in the vertical as indicated in table 7. Keep equipment several feet from piers and abutments if velocities are high. Estimate the depth and velocity next to the pier or abutment on the basis of the observations at the nearest vertical.

If there are piers in the cross section, it is usually necessary to use more than 25 to 30 partial sections to get results as reliable as those from a similar section without piers. Piers will often cause horizontal angles that must be carefully measured. Piers also cause rapid changes in the horizontal-velocity distribution in the section.

Whether or not to exclude the area of a bridge pier from the area of the measurement cross section depends primarily on the relative locations of the measurement section and the end of the pier. If measurements are made from the upstream side of the bridge, it is the relative location of the upstream end (nose) of the pier that is relevant; for measurements made from the downstream side, it is the location of the downstream end (tail) of the pier that is relevant. If any part of the pier extends into the measurement cross section, the area of the pier is excluded. Bridges quite commonly have cantilevered walkways from which discharge measurements are made. In these cases, the measurement cross section lies beyond the end of the pier (upstream from the pier nose or downstream from the pier tail, depending on which side of the bridge is used). In that situation, it is the position and direction of the streamlines that determines whether or not the pier area is to be excluded. If the stationing of the sides of the pier when projected to the measurement cross section was not already done, the hydrographer does it at this point. If there is negligible or no downstream flow in that width interval (pier subsection), then the pier is excluded. That is, if only stagnation and (or) eddying exists upstream from the pier nose or downstream from the pier tail, whichever is relevant, the area of the pier is excluded. If there is substantial downstream flow in the pier subsection, the area of the pier is included in the area of the measurement cross section. In that circumstance, the horizontal angles of the streamlines in and near the pier subsection will usually be quite large.

Footbridges are sometimes used for measuring canals, tailraces, and small streams. Rod suspension can be used from many footbridges. The procedure for determining depth in low velocities is the same as for wading measurements. For higher velocities, obtain the depth by the difference in readings at an index point on the bridge when the base plate of the rod is at the water surface and on the streambed. Measuring the depth in this manner will eliminate errors caused by the water piling up on the upstream face of the rod. ADCP tethered boats, handlines, bridge cranes, and bridge boards are also used from footbridges.

The handline can be disconnected from the headphone wire and passed around a truss member with the sounding weight on the bottom. This eliminates the need for raising the weight and meter to the bridge each time a move is made from one vertical to another; it is the principal advantage of a handline.

Safety is a primary consideration when measuring discharge from bridges. High-speed traffic can present a major safety hazard; in fact, it is no longer permissible to make discharge measurements from some Interstate route bridges without special permission. Observe all safety precautions, such as the use of traffic cones, traffic signs, and flag persons, that are prescribed in the USGS Water Science Center's flood plan and safety plan.

Observe the same safety precautions regarding the snagging of debris, such as floating or submerged trees or logs, as described above for cableways.

Current-Meter Measurements From Ice Cover

Discharge measurements under ice cover, as shown in figure 18, are made under the most severe conditions, but are extremely important because a large part of the discharge record during a winter period may depend on one measurement. In recent years ADCPs and ADVs have increasingly been used to make measurements from ice cover.

Select the possible locations of the cross section to be used for measurement from ice cover during the open-water season when channel conditions can be evaluated. Commonly, the most desirable measurement section will be just upstream from a riffle because slush ice that collects under the ice cover is usually thickest at the upstream end of the pools created by riffles.

The equipment used for cutting or drilling the holes in the ice is described in a previous section of this chapter.

Never underestimate the danger of working on ice-covered streams. When crossing, test the strength of the ice with solid blows using a sharp ice chisel. Ice thickness may be irregular, especially late in the season when a thick snow cover may act as an insulator. Water just above freezing can slowly melt the underside of the ice, creating thin spots. Ice that is bridged above the water may be thick but still be weak.

Cut the first three holes in the selected cross section at the quarter points to detect the presence of slush ice or poor distribution of the flow in the measuring section. If poor conditions are found, investigate other sections to find one that is free of slush ice and that has good distribution of flow. After finding a suitable cross section, make at least 20 holes in the ice for a current-meter measurement. Space the holes so that no partial section contains more than 10 percent of the total discharge. On narrow streams, it may be simpler to remove all of the ice in the cross section.

The effective depth of the water, as shown in figure 19, is the total depth of water minus the distance from the water surface to the bottom of the ice. The vertical pulsation of water

Figure 18. *A*, Ice drill being used to cut holes and *B*, ice rod being used to support current meter for a discharge measurement.

in the holes in the ice sometimes causes difficulty in determining the depths. The total depth of water is usually measured with an ice rod or with a sounding weight and reel, depending on the depth.

Measure the distance from the water surface to the bottom of the ice with an ice-measuring stick. Do not use the ice-measuring stick if there is slush under the solid ice at a hole. In order to find the depth at which the slush ice ends, suspend the current meter below the slush ice with the meter rotor turning freely. Raise the meter slowly until the rotor stops. This point is used as the depth of the interface between water and slush. After the effective depth of the water has been determined, compute the proper position of the meter in the vertical as shown in figure 19.

Use the Price winter Water Survey of Canada (WSCan) current meter yoke, with a polymer rotor, under ice cover when slush ice is present because the cups are solid and cannot become filled with slush ice; this is what happens with the cups of the regular Price meter. For situations where slush ice is not present, use the Price winter WSCan current meter yoke

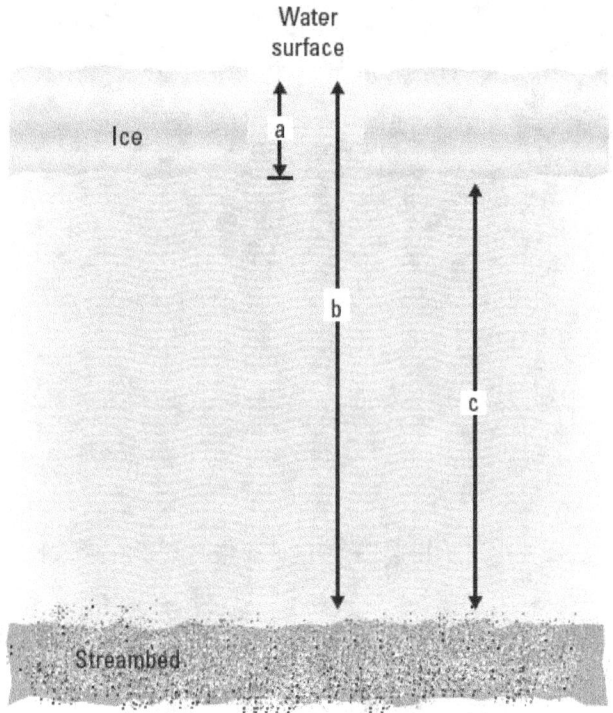

a = Water surface to bottom of ice   0.2-depth setting = a + 0.2c
b = Total depth of water   0.8-depth setting = b - 0.2c
c = Effective depth (c=b-a)   0.6-depth setting = b - 0.4c

Figure 19. Method of computing meter settings for measurements under ice cover.

with regular Price metal cups. The old-style vane ice meter is no longer recommended, primarily because of its poor performance in slow velocities.

The velocity distribution under ice cover, when the water is in contact with the underside of the ice, is similar to that in a pipe, with a lower velocity nearer the underside of the ice. This is illustrated in figure 20. Use the 0.2- and 0.8-depth method for effective depths of 2.5 ft or greater, and the 0.6-depth method for effective depths of less than 2.5 ft. Define two vertical-velocity curves while making ice measurements to determine whether any coefficients are necessary to convert the velocity (obtained by the 0.2- and 0.8-depth method or the 0.6-depth method) to the mean velocity. Normally, the average of the velocities obtained by the 0.2- and 0.8-depth method gives the mean velocity, but a coefficient of about 0.92 usually is applicable to the velocity obtained by the 0.6-depth method.

When measuring the velocity, keep the meter as far upstream as possible to avoid any effect that the vertical pulsation of water in the hole might have on the meter. Eliminate as much as possible the exposure of the meter to the cold air during the measurement. The meter must be free of ice when the velocity is being measured.

If there is partial ice cover at a cross section, use the procedure described above where there is ice cover, and use open-water methods elsewhere.

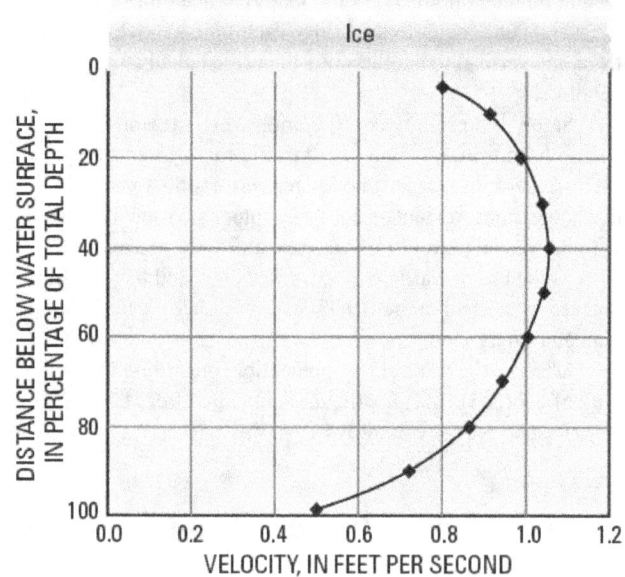

Figure 20. Typical vertical-velocity curve under ice cover.

DISCHARGE MEASUREMENT NOTES—ICE COVER
FRIO   River, at   COLD HARBOR, ME
Creek, near
METER = V-587

REW Dist. from initial point	Width	Total depth of water	W.S. to bot. ice	Effective depth	Depth of meter below water surface	Revolutions	Time in seconds	VELOCITY At point	VELOCITY Mean in vertical	Area	Discharge
1315											
24	3	0							0	0	0
30	6	2.6	1.6	1.0	2.2	10	47	.49	.45	6.0	2.7
36	6	3.6	1.8	1.8	2.9	15	49	.69	.63	10.8	6.8
42	5	4.3	2.0	2.3	3.4	15	44	.77	.71	11.5	8.2
46	4	4.5	1.8	2.7	2.3	20	43	1.04	.84	10.8	9.1
					4.0	15	52	.65			
50	4	4.7	1.7	3.0	2.3	20	40	1.12	.90	12.0	10.8
					4.1	15	50	.68			
54	4	4.6	1.7	2.9	2.3	25	49	1.14	.94	11.6	10.9
1325					4.0	15	46	.74			
58	4	4.9	1.6	3.3	2.3	20	40	1.12	.90	13.2	11.9
					4.2	15	50	.68			
62	4	4.8	1.6	3.2	2.2	25	48	1.17	.96	12.8	12.3
					4.2	15	46	.74			
66	3.5	5.0	1.5	3.5	2.2	25	44	1.27	1.04	12.2	12.7
					4.3	15	41	.82			
69	3	5.3	1.6	3.7	2.3	25	40	1.40	1.12	11.1	12.4
					4.6	15	40	.84			
72	3	5.1	1.5	3.6	2.2	25	41	1.36	1.12	10.8	12.1
1335					4.4	20	51	.88			
	49.5									122.8	109.9

Figure 21. Part of note sheet for discharge measurement under ice cover.

A sample sheet of discharge-measurement notes under ice cover is shown in figure 21. In this measurement, the vertical-velocity curves indicate that the 0.2- and 0.8-depth method gives the mean velocity and that the 0.6-depth method requires a coefficient of 0.92.

Current-meter measurements under ice cover are frequently made with a special winter-style sounding rod, ADV or ADCP, as described in this chapter. When depths are too deep for rod suspension, use an equipment assembly mounted on runners, such as shown in figure 22. For winter conditions, the 30-pound C-type weights should be used with a special, collapsible hanger assembly (shown in figure 22) that can be passed through an 8-in. hole in the ice. A handline can also be used for making ice measurements.

Where it is impractical to use a powered ice drill, use ice chisels to cut the holes. Ice chisels are usually 4- or 4.5-ft long and weigh about 12 pounds. Use the ice chisel when first crossing an ice-covered stream to determine whether the ice is strong enough to support the hydrographer. If a solid blow of the chisel blade does not penetrate the ice, it is safe to walk on, providing the ice is in contact with the water.

Some hydrographers supplement the ice chisel with a Swedish ice auger. The cutting blade of this auger is a spade-like tool of hardened steel that can cut a hole 6 to 8 in. in diameter by turning a brace-like arrangement on top of the shaft.

After the hole is made in the ice, water will be forced up, owing to the water being under pressure from the weight of the ice. In order to determine the effective depth of the stream, use ice-measuring sticks to measure the distance from the water surface to the bottom of the ice. Measuring this distance is done using a bar about 4 ft long, made of strap steel or wood, graduated in feet and tenths of a foot and having an L-shaped projection at the lower end. Hold the horizontal part of the L on the underside of the ice and read the depth to that point at the water surface on the graduated part of the stick. The horizontal part of the L is at least 4 in. long so that it may extend beyond any irregularities on the underside of the ice.

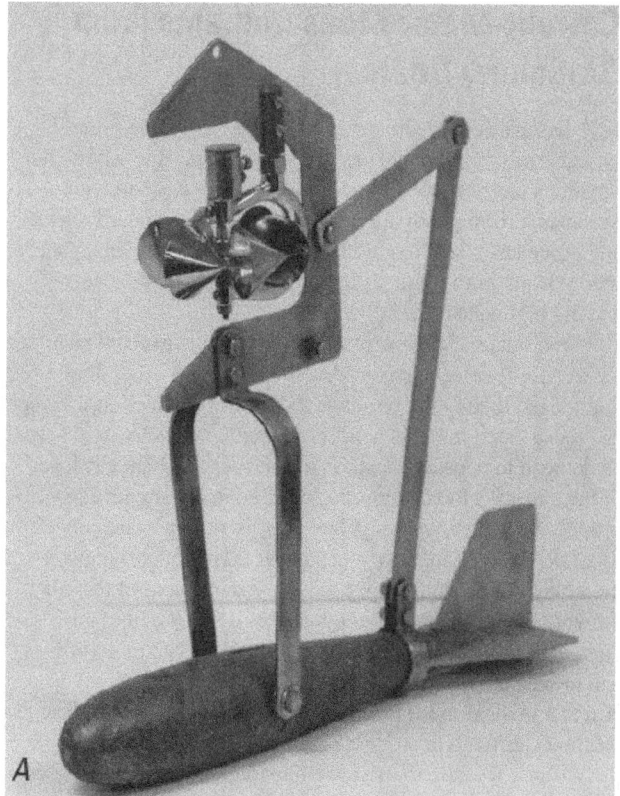

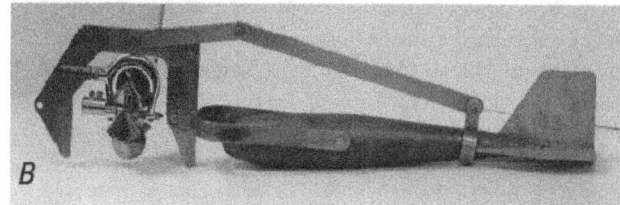

Figure 22.   Collapsible hanger assembly for used with 30- and 50-pound C-type weights, for measurements under ice (*A*, in measurement position and *B*, collapsed).

Current-Meter Measurements From Stationary Boats

Discharge measurements are made from boats where no cableways or suitable bridges are available and where the stream is too deep to wade, although ADCP discharge measurements from a moving boat, now a USGS standard operating procedure, have largely replaced this method. Personal safety is the limiting factor in the use of boats on streams having high velocity of flow.

For boat measurements whether using a mechanical meter or an ADCP, select a cross section that has attributes similar to those described in the previous section "Site selection," except for those listed in items concerning depth and velocity. There is no need to consider depth in a boat measurement because if the stream is too shallow to float a boat, the stream can usually be waded. Velocity, however, is an important concern. If velocities are too slow, mechanical current meter registration may be affected by an oscillatory movement of the boat, in which the boat (even though fastened to a tag line) moves upstream and downstream as a result of wind action. Vertical movement of the boat as a result of wave action may also affect a vertical-axis current meter. If velocities are too fast, it becomes difficult to string a tag line across the stream.

If it is feasible to use a tag line in making a boat measurement, string it at the measuring section by unreeling the line as the boat moves across the stream. After a tag line without a brake has been stretched across the stream, take up the slack by means of a block and tackle attached to the reel and to an anchored support on the bank. If there is traffic on the river, one person must be stationed on the bank to lower and raise the tag line to allow the river traffic to pass. Place streamers on the tag line so that it is visible to boat pilots. If there is a continual flow of traffic on the river, or if the width of the river is too great to stretch a tag line, other means will be needed to position the boat. Night measurements by boat are not recommended because of the safety concerns.

When no tag line is used, the boat can be kept in the cross section by lining it up with flags positioned on each end of the cross section, as illustrated in figure 23. Flags on one bank would suffice but it is better to have them on both banks. Determine the position of the boat in the cross section by using a transit or total station on the shore and a stadia rod held in the boat. Another method of determining the position of the boat is by setting a transit or total station on one bank a convenient, known distance from and at right angles to the cross-section line. The position of the boat is determined by measuring the angle α to the boat, measuring the distance CE, and computing the distance MC as shown in figure 24. A third method of determining the position of the boat is done with a sextant read from the boat. Position a flag on the cross-section line and another at a known distance perpendicular to the line. The boat position can be computed by measuring the angle β with the sextant, as shown in figure 24. Boat position can also be determined by using a global positioning system with differential corrections (DGPS). This method is especially useful on wide streams and in flood plains where other methods of determining boat position are not applicable. Unless anchoring is more convenient, the motor must hold the boat stationary while readings are being taken.

Do not take boat measurements at velocities less than 1 ft/s when the boat is subject to wave action. The up-and-down movement of the boat (and the meter) seriously affects the velocity observations. If the maximum depth in the cross section is less than 10 ft and the velocity is low, the hydrographer can use a rod for measuring the depth and for supporting the current meter. For greater depths and velocities, use a cable suspension with a reel and sounding weight. The procedure for measuring from a stationary boat using the boat boom and crosspiece is the same as that for measuring from a bridge or a cableway, as described in previous sections of this chapter.

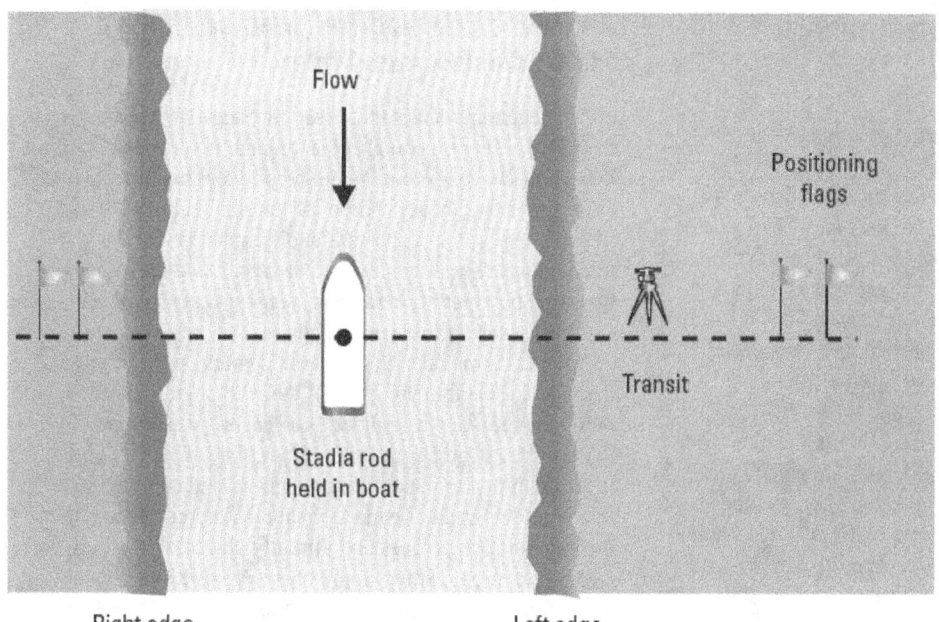

Flow

Positioning
flags

Transit

Stadia rod
held in boat

Right edge
of water

Left edge
of water

Figure 23. Determining the
position in a cross section
using the stadia method.

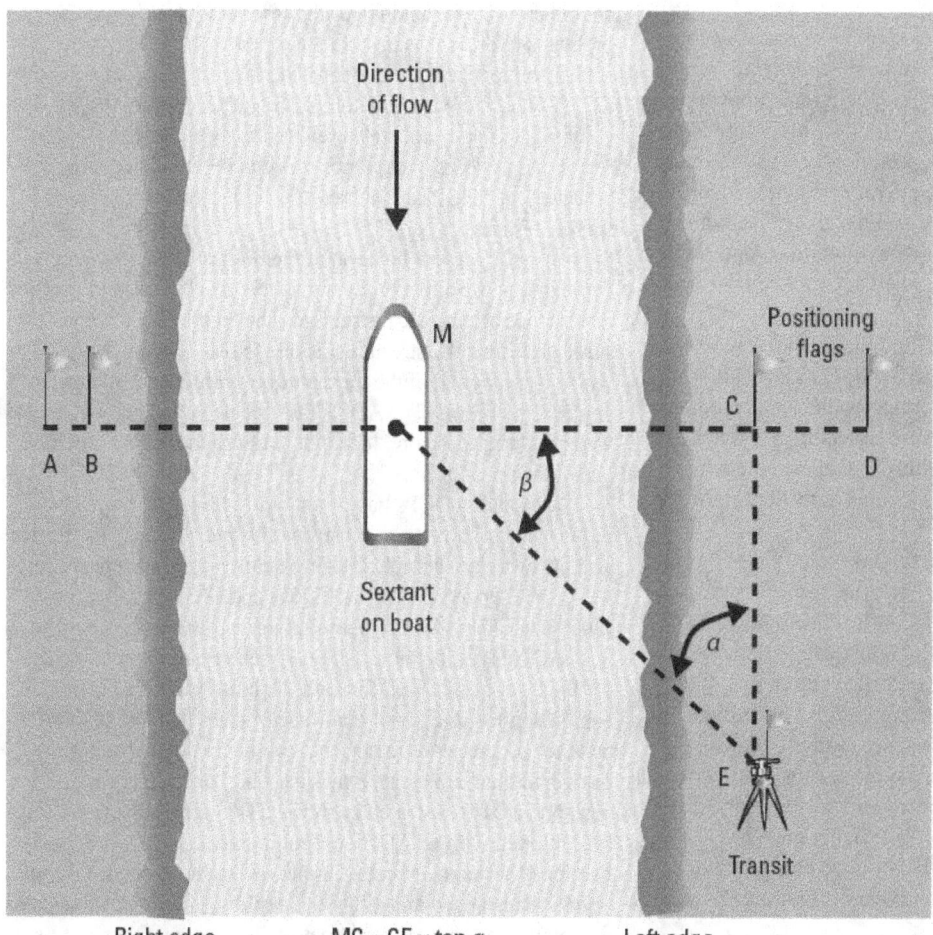

Direction
of flow

M

Positioning
flags

C

A B

β

D

Sextant
on boat

α

E

Transit

Right edge
of water

$MC = CE \times \tan \alpha$
$MC = CE \div \tan \beta$

Left edge
of water

Figure 24. Determining the
position in a cross section
using the angular method.

Moving-Boat Measurements of Discharge

On large streams and estuaries, the midsection methods of measuring discharge are frequently impractical and involve costly and tedious procedures. There may be no facilities at remote sites. Where facilities do exist, they may be inundated or inaccessible during floods. At some sites, unsteady flow conditions require that measurements be made as rapidly as possible. Measurements on tide-affected rivers must not only be made frequently, but continually, throughout a tidal cycle. The moving-boat technique is a method of quickly measuring a large stream. It requires no fixed facilities, and it lends itself to the use of alternate sites if conditions make this applicable. Brief descriptions of three moving-boat methods are given in the following sections. It must be stated here that these methods have been almost entirely replaced by the use of ADCP discharge measurements. Details of ADCP setup and use can be found in Mueller and Wagner (2009) and subsequent section of this chapter discussing the moving boat method using an ADCP.

Manual Method, Using a Mechanical Current Meter

Smoot and Novak (1968) describe the manual moving-boat technique in detail. It is similar to the mechanical current-meter measurement in that the velocity-area approach is used to determine the discharge. The total discharge is the summation of the products of the partial areas of the stream cross section and their respective average velocities. During the traverse of a stream by boat, a sonic sounder records the geometry of the cross section, and a continuously operating current meter senses the combined stream and boat velocities.

The current meter used for a moving-boat measurement is a component propeller type, such as the Ott current meter, with a custom body made for mounting on the leading edge of a vane. The current meter and vane assembly are attached to a vertical rod and bearing assembly that allows them to rotate freely. An angle indicator is located at the top of the rod, which indicates the crab angle.

It usually takes three people to make a moving-boat measurement: one to operate the boat, one to make crab-angle observations, and one to take notes. The note keeper is responsible for recording the crab-angle observations and the current-meter pulses. Depths are recorded automatically on the sonic-sounder chart. Special computations and conversions, which are somewhat tedious, are required to determine stationing and normal mean velocities for each vertical. Discharge can then be computed similar to a standard velocity-area discharge measurement. Experience has shown that measurements obtained by the moving-boat technique compare within 5 percent of measurements obtained by conventional means.

Automatic Computerized Method, Using a Mechanical Current Meter

The automatic computerized moving-boat method is the same basic procedure as the manual moving-boat method described above, except that all readings of depth, velocity, crab angles, and boat position are automatically fed into an onboard computer. A substantial difference is that the automated moving-boat method requires only two crew members—a boat operator and an instrument operator. The manual method requires three crew members.

The automated method requires an electronic compass that provides automatic input of the crab angle. The current meter and depth sounder likewise provide automatic input of pulse rates and depth, respectively. The computer is programmed to make the conversions and computations that would normally be made manually, thereby speeding up the computation process and virtually eliminating arithmetic errors. A computed measurement is available immediately after completion of the data collection.

Moving-Boat Method, Using an ADCP

ADCPs can be used to measure unsteady, bidirectional, and other flows with nonlogarithmic velocity distributions—problems hydrologists have faced for decades. ADCPs are called profilers because they provide measurements of velocity throughout the water column. The ADCP divides the water column into depth cells (also referred to by some software and references as "bins") and reports a velocity for each depth cell; however, an ADCP cannot measure velocities near the water surface or near the bed. The length of the unmeasured zone at the water surface is a function of the draft of the instrument deployment, the effect of the transducer mechanics, and the flow disturbance around the instrument. The length of the unmeasured zone near the streambed is due to side-lobe interference, which is a function of the mechanics of transducers and the slant angle of the beams. The ADCP must be deployed below the water surface; therefore, it cannot measure the water velocity above the transducers.

Although ADCPs have no moving parts and typically require no calibration, the instruments and associated software and firmware are complex. Using quality-assurance procedures defined by Oberg and others (2005) and Mueller and Wagner (2009) will help identify potential instrument problems.

ADCPs can be mounted on either side of manned boats, off the bow, or in a well through the hull. A tethered boat can be defined as a small boat (usually less than 6.5 ft long) attached to a rope, or tether, that can be deployed from a bridge, a fixed cableway, a moving boat, or a temporary or permanent bank-operated cableway. The tethered boat should be equipped with an ADCP mount that meets all of the specifications for manned boats. Unmanned, remote-control ADCP boats are also used and allow the deployment of ADCPs where deployment with a manned boat or tethered boat may not be feasible or ideal. Similar to (but smaller than) a manned boat, a remote-control boat has self-contained motors and a remote-control system for maneuvering the boat across the river.

The procedures for predeployment preparation, field data collection, and processing of collected data are discussed in detail by Mueller and Wagner (2009). A detailed description of how an ADCP measures velocity and computes discharge and additional details on selected topics are presented in the appendices (Mueller and Wagner, 2009).

Networks of Current Meters

In the past, occasional special measurements made by USGS hydrographers have required simultaneous velocities at several points in a cross section, distributed either laterally or vertically. For example, it may be necessary to measure a vertical-velocity profile quickly in unsteady flows and to check it frequently in order to determine the changes in shape of the vertical profile, as well as the rates of these changes. In another example, for the measurement of tide-affected streams, it is desirable to measure the total discharge continuously during at least a full tidal cycle (approximately 13 hours). The need for so many simultaneous velocity determinations (one at each vertical in the cross section) for so long a period could be an expensive and laborious process using conventional techniques of discharge measurement.

A grouping of 21 current meters and special instrumentation has been devised in the past by the USGS to facilitate measurements of the types just described. The 21 meters are connected together so that the spacing between any two adjacent meters can vary up to 200 ft. In using this method, each meter should be uniformly calibrated and have sufficient handline cable to be suspended vertically from a bridge as much as 200 ft. Revolutions of the rotors are recorded by electronic counters that are grouped compactly in one box at the center of the bank of meters. The operator, by flipping one switch, starts all 21 counters simultaneously, and after an interval of several minutes, stops all counters. The indicated number of revolutions for the elapsed time interval is converted to a velocity for each meter. The distance between meters is known, and a record of stage is maintained to evaluate depth. Prior information at the site is obtained to convert point velocities in the verticals to mean velocities in those verticals. All of the information necessary to compute discharge in the cross section is available, and is tabulated for easy conversion to discharge. If possible a concurrent ADCP discharge measurement should be completed in close proximity and time to this measurement to corroborate the results.

Discharge Measurement of Deep, Swift Streams With a Mechanical Current Meter

Discharge measurements of deep, swift streams with a mechanical current meter usually present no serious problems when adequate sounding weights are used and when floating drift or ice is not excessive. Normal procedures must sometimes be altered, however, when measuring these streams. The four most common circumstances are the following:

1. It is possible to sound, but the weight and meter drift downstream.

2. It is not possible to sound, but a standard cross section is available.

3. It is not possible to sound, and a standard cross section is not available.

4. It is not possible to put the weight and meter in the water.

Procedures are described below for use during measurements made under each of these conditions. Use procedures for items 2, 3, and 4 where there is a stable cross section. The procedure for unstable channels must be determined by conditions at each location.

Possible To Sound, but Weight and Meter Drift Downstream

For some streams, it may be possible to sound the streambed, but because of the force of high velocities, the weight and meter are carried downstream. This may be a condition for only a few verticals near the center of the stream, or it may affect many of the verticals. Make corrections to the observed depths and meter settings to account for the downstream drift. These corrections are commonly referred to as "vertical angle corrections." The procedure for computing vertical angle corrections is described in a previous section of this chapter entitled "Depth corrections for downstream drift of current meter and weight." The corrections can be computed manually, or they may be computed automatically through the use of an electronic notebook or a PDA.

Not Possible To Sound, but Standard Cross Section Available

When it is not possible to sound the streambed, use a standard cross section from previous measurements at the bridge or cableway for determining depth. Such a cross section is useful only if all discharge measurements use the same permanent initial point for the stationing of verticals across the width of the stream. There should also be an outside reference gage or reference point on the bank or bridge to which the water-surface elevation at the measurement cross section may be referred. If these conditions are met, use the following procedure to make a discharge measurement:

1. Determine the depths from the standard cross section, based on the water-surface elevation.

2. Measure the velocity at 0.2 of the depth at each vertical.

3. Compute the measurement in the normal manner using the measured 0.2-depth velocities as though they were the mean velocities in the vertical. Apply horizontal-angle corrections, if necessary. Use depths as determined in step 1 above.

4. Determine the coefficient to adjust the 0.2-depth velocity to the mean velocity on the basis of previous measurements at the site by the two-point method. See a previous section of this chapter entitled "Two-tenths-depth method."

5. Apply the coefficient from step 4 to the computed discharge from step 3.

Not Possible To Sound, and Standard Cross Section Not Available

When it is not possible to sound the streambed and a standard cross section is not available, use the following procedure:

1. Reference the water-surface elevation before and after the measurement to an elevation reference point on a bridge, on a driven stake, or on a tree at the water's edge. It is assumed here that no outside reference gage is available at the measurement cross section.

2. Estimate the depth and observe the velocity at 0.2 of the estimated depth. The meter should be at least 2.0 ft below the water surface. In the notes, record the actual depth the meter was placed below the water surface. If an estimate of the depth is impossible, place the meter 2.0 ft below the water surface and observe the velocity at that point.

3. Make a complete measurement, including some vertical-velocity curves, at a lower stage when you can sound the streambed.

4. Use the complete measurement and difference in stage between the two measurements to determine the cross section of the first measurement. To determine whether the streambed has shifted, compare the cross section with one taken for a previous measurement at that site.

5. Use vertical-velocity curves, or the relation between mean velocity and 0.2-depth velocity, to adjust the velocities

observed in step 2 to mean velocity. Apply horizontal-angle corrections as necessary.

6. Compute the measurement in the normal manner using the depths from step 4 and the velocities from step 5.

Not Possible To Put the Weight and Meter in Water

If it is impossible to put the sounding weight and mechanical current meter in the water because of high velocities and (or) floating drift, use the following procedure:

1. Obtain depths at the measurement verticals from a standard cross section, if one is available. If a standard cross section is not available, determine depths by the method explained above in the section "Not possible to sound, and standard cross section not available."

2. Measure velocities and compute discharge using an ADCP.

3. Measure surface velocities by timing floating drift, or by using an optical or other approved noncontact flowmeter.

4. Compute the measurement in the normal manner, using the surface velocities as though they were the mean velocities in the vertical, and using the depths from step 1.

5. Apply the appropriate velocity coefficient to the discharge computed in step 3. Use a coefficient of 0.86 for a natural channel and 0.90 for an artificial channel.

The optical current-meter and ADCP measurement procedures are described in previous sections of this chapter and by Mueller and Wagner (2009). The optical current meter is portable, battery operated, and requires no great skill for quick and accurate readings of the surface rate of flow. The meter is not immersed, so it does not disturb the flow, and it is in no danger of damage from floating debris or ice. In many cases, the ADCP has become the most efficient alternative to the mechanical current meter where velocities are too great; however, the ADCP also has limitations. See the previous sections of this chapter for information on velocity, depth, turbidity, and other site-condition limitations of the ADCP.

Keep in mind that just after the crest, the amount of floating drift or ice is usually greatly reduced, and it may be possible to obtain velocity observations with a current meter. These observations can help define the velocity coefficient mentioned in step 5 above.

Recording Field Notes

Field notes for a discharge measurement may be recorded on standard paper note sheets (for example, USGS Forms 9-275-F, 9-275-I, and other special field forms). With the ADCP discharge measurement, the software attached to each instrument contains digital forms for the recording of some of the field data. The USGS has developed a paper form for recording field data observed during an ADCP discharge measurement (fig. 2D). With a current-meter discharge measurement, field forms can be recorded using an electronic notebook, such as the Aquacalc or a Personal Digital Assistant (PDA). With an ADV measurement, there are special field forms to accommodate its specifications and details. These methods are described in more detail in subsequent paragraphs in this section. The SWAMI program with a PDA (commonly used by the USGS) can be used to record discharge measurements, inspections, differential level surveys, and other field measurements. SWAMI has an interface with the National Water Information System (NWIS), so measurements are easily uploaded to NWIS (fig. 2C).

Standard Paper Note Keeping for a Mechanical Current-Meter Discharge Measurement

Paper note sheets, as shown in figure 2A, are the traditional way to record the field observations for a mechanical current meter, ADV, or ADCP discharge measurement. Generally, for each discharge measurement, the hydrographer should record the following information, at a minimum, on the front sheet of the measurement notes (the information may vary, depending on the meter and method being used):

- Measurement number, who computed, and who checked the measurement;
- Downstream station identification number and station name (station name includes stream name and location, to correctly identify an established gaging station). For a miscellaneous measurement, record the stream name and exact location of site;
- Date of measurement and members of measurement party (initials and last name);
- Measured channel width, area, average velocity (computed as a ratio of the measured discharge/measured area), average gage height, and discharge;
- Vertical velocity method(s) of measurement, number of sections, and change in gage height during the discharge measurement;
- Measurement method coefficient, horizontal-angle coefficient, type of meter suspension (for example, rod, 100#C, and so forth) and whether tags were checked;
- Type of meter (for example, AA or pygmy), the current meter's serial number; and the elevation of the meter above the channel bottom;

- Meter rating used (for example, Standard Rating No. 2) and the most recent spin test results;
- Measurement percentage (after computed) from the existing stage-discharge rating, and the indicated shift in feet from that rating;
- GAGE READINGS: Do not erase inside this block on the front sheet. If an error is made, cross through the error and write the correct reading.
  - Start time measurement using 24-hour clock time, and record the time zone (that is, EST, CST, EDT, and so forth).
  - Record inside and outside gage, and also readings from recording devices (for example, data logger, graphic, and so forth).
  - Compute weighted mean gage height either by averaging readings, or if sufficient change in gage height occurred, by using methods for weighting gage height discussed in this chapter.
  - Compute gage-height correction caused by difference in true gage height (reference gage) and recorder or other gage that is reading incorrectly.
  - Record the correct mean gage height.
- Samples collected: Indicate type of water-quality measurements and samples [that is, water-quality, sediment, and (or) biological], and indicate if the measurements are documented on separate sheets (that is, water quality, aux./base gage, other);
- Indicate whether the rain gage (if applicable) was serviced/calibrated;
- Briefly describe the weather (for example, sunny, cloudy, rainy, cold, or other);
- Record the air temperature in degrees Celsius and the time of the reading;
- Record the water temperature in degrees Celsius and the time of the reading;
- Record the check bar reading (if a wire weight is present), time of the reading, and any adjustments in elevation made to the check bar.
- Indicate the type of measurement (wading, cable, ice boat, and so forth) and location of measurement relative to the gage (upstream, downstream, and so forth).
- Rate the measurement based on the hydrologic/hydraulic conditions in which the measurement was made [that is, excellent (2 percent), good (5 percent), fair (8 percent), or poor (more than 8 percent)].
- Flow: Document the hydraulic condition of the flow (steady, unsteady, where the flow was within the cross section, and so forth).
- Cross section: Geomorphologically describe the cross section (that is, sand, clay, cobble, and so forth), shape, presence of vegetation, and any other roughness affecting flow.

- Document if the gage is operating and whether the record was removed during this visit;

- Note the battery voltage and the cleaning of the orifice or intakes;

- If appropriate, indicate pressure readings of the nonsubmersible pressure transducer (that is tank, line, bubble-rate (bubbles/minute);

- Indicate if appropriate, readings of extreme indicators of high flow.

- Document the condition of the crest stage gage (CSG) and record the high water mark (HWM) if a reading is available on the CSG and the reference elevation.

- Record any other HWM obtained at the gage, if appropriate.

- Control conditions: Describe what and where the control of flow is for the gage pool (that is, gravel riffle about 80 ft downstream of the gage, and so forth).

- REMARKS: Use this space to document any unusual conditions in the gage reach that might affect the measurement, record, or other pertinent information regarding the accuracy of the discharge measurement and conditions that might affect the stage-discharge relation and document any observer reading or results of discussions with an observer.

- Measure the gage height of zero flow as many times a year as possible; record to the nearest 0.1 ft. Do this by recording the gage height at the gage at the time the gage height at zero flow is measured. Subtract the gage height from the depth of flow at the point the gage height at zero flow is measured and rate this measurement as good, fair, or poor.

- Fill in all items on the front sheet or mark with a dash after the measurement is completed and computed.

Mechanical Current-Meter Inside Notes

In the inside notes of a mechanical current-meter discharge measurement, identify the measurement starting point by either left edge of water or right edge of water (LEW or REW, respectively), when facing downstream, and record the time you started the measurement. If a significant change in stage is expected during the measurement, periodically record the time for intermediate verticals during the course of the measurement. If possible, synchronize this time with the recording interval of the digital recorder or data logger. Intermediate times are important because if there is any appreciable change in stage during the measurement, these recorded times are used to determine intermediate gage heights, which are then used to compute a weighted mean gage height for the measurement, as described in a subsequent section of this chapter. When the measurement is completed, record the time and the bank of the stream (LEW or REW) where the section ends.

Begin the measurement by recording the distance from the initial point to the edge of the water. Measure and record the depth, and velocity (if any), at the edge of water. Compute the width using the midsection method described in a previous section of this chapter. Proceed across the measurement section by measuring and recording the distance of each vertical from the initial point; the depth at the vertical; the observation depths as 0.6, 0.2, 0.8, and so forth; the revolutions and time for each velocity observation; and the horizontal angle coefficient if different than 1.00.

Complete all computations required for the inside notes to determine the total width, area, and discharge. Transfer these values to the front sheet and complete other items on the measurement front sheet. The measurement computations should be made, and the note sheets completed, before the hydrographer leaves the gaging station.

Erasures of original field data are not allowed. This includes items such as gage readings, distances, depths, meter revolutions, times, horizontal-angle coefficients, and other field measurements that cannot be repeated. If a variable is remeasured, and it is necessary to change the originally recorded value of that variable, cross it out and record the new measurement above or adjacent to the original. The original measurement should remain legible, even though it is crossed out. On the other hand, it is permissible to erase computed values, such as velocities, areas, widths, and discharges.

Standard Paper Note Keeping for an ADV Discharge Measurement

Paper note sheets, as shown in figure 2B, are the traditional method to record the field observations for an ADV discharge measurement. If using paper note keeping with an ADV for each measurement, the hydrographer should record the information, at a minimum, as if it were a mechanical velocity-meter discharge measurement, with a few variations. Indicate the filename of the infield diagnostic test performed on the ADV during the discharge measurement. Fill in all items on the front sheet or mark with a dash after the measurement is completed and computed.

ADV Inside Notes

An electronic summation of an ADV discharge measurement is produced by the ADV software. This information contains much of the information on the front sheet of the standard discharge measurement form, plus depths, widths, velocities, angles, area, and discharge. Typically ADV software does all the computations for an ADV discharge measurement. Print this output and attach it to the discharge measurement form for archival. Recent programming with personal digital assistants (PDAs) has further facilitated the collection, processing, and entry of discharge measurements into digital databases. See the sections entitled Electronic counters" and "Other electronic counters, electronic field notebooks, and personal digital assistants" for further discussion of the use of PDAs for field measurements.

Mean Gage Height of Discharge Measurements

The mean gage height of a discharge measurement represents the mean height of the stream during the period the measurement was made and is referenced to the datum of the gaging station. Just as an accurate determination of the discharge is important, so is an accurate determination of the mean gage height because it is one of the coordinates used in plotting the discharge measurement to establish the stage-discharge relation. The computation of the mean gage height presents no problem when the change in stage is small (0.1 ft or less). At gage-height changes of less than 0.1 ft, the mean gage height for the discharge measurement can be obtained by averaging the gage heights at the beginning and end of the measurement, without significant error. Measurements, however, must sometimes be made during floods or regulation when the stage significantly changes more than 0.1 ft.

To compute an accurate mean gage height for a discharge measurement, read the gage at the beginning and end of the discharge measurement, and several times during the measurement if there are significant changes. If the station is equipped with an electronic data logger or DCP that automatically records at intervals of 15 minutes or less, you can take the intermediate gage-height readings from those instruments after the measurement is completed. The hydrographer should accurately synchronize watch time and recorder time, and should record watch time for selected verticals at intervals during the discharge measurement. If the recording interval is greater than 15 minutes (that is, 30 minutes or 1 hour), intermediate gage- height readings should be obtained by reading the gage once or twice during the discharge measurement.

If the change in stage during the measurement is greater than about 0.1 ft (Rantz, 1982, suggests a change of 0.15 ft), the mean gage height should be computed by weighting the gage-height readings. In the past, the mean gage height was computed by weighting the gage readings with partial discharges from the discharge measurement. Later studies show, however, that this method tends to overestimate the mean gage height. Time weighting has also been used to compute a weighted mean gage height, but this method tends to underestimate the mean gage height. Therefore, it is recommended that both methods of weighting be used for discharge measurements having stage changes of 0.10 or more, and that an average of the two results be used for the mean gage height.

Plot the gage-height readings so that intermediate readings can be interpolated where necessary. Pay particular attention to breaks in the slope of the gage-height graph. Figure 25 illustrates a plot of gage heights for a discharge measurement. Gage heights for this measurement were determined from the stage recorder at 15-minute intervals.

In the discharge-weighting procedure, the partial discharges measured between recorded watch times are used with the mean gage height for that same time period. The equation used to compute the weighted mean gage height is:

$$H = \frac{q_1 h_1 + q_2 h_2 + q_3 h_3 + ... + q_n h_n}{Q}, \quad (11)$$

where
H		weighted mean gage height, in feet,
Q		total discharge measured, in cubic feet per second $= q_1 + q_2 + q_3 + ... + q_n$,
$q_1, q_2, q_3, ... q_n$		amount of discharge measured during time interval 1, 2, 3...n, in cubic feet per second,
$h_1, h_2, h_3, ... h_n$		average gage height during time interval 1, 2, 3...n, in feet.

Figure 25 shows the computation of a discharge-weighted mean gage height. The graph at the bottom is a reproduction of the gage-height graph during the discharge measurement. To help explain the method, the discharges are taken from the current-meter measurement shown in figure 2A. The upper computation of the mean gage height in figure 25 shows the computation using the given formula. The lower computation was calculated using a shortcut method to eliminate the multiplication of large numbers. In this method, after the average gage height for each time interval has been computed, a base gage height, which is usually equal to the lowest average gage height, is chosen. Then, the difference between the base gage height and the average gage heights is used to weight the discharges. When the mean difference has been computed, the base gage height is added to it.

In the time-weighting procedure, the arithmetic mean gage height for time intervals between breaks in the slope of the gage-height graph is used with the duration of those time periods. The equation used to compute mean gage height is

$$H = \frac{t_1 h_1 + t_2 h_2 + t_3 h_3 + ... + t_n h_n}{T}, \quad (12)$$

where
H		weighted mean gage height, in feet,
T		total time for the measurement, in minutes $(t_1 + t_2 + t_3 + ... + t_n)$,
$t_1, t_2, t_3, ..., t_n$		duration of time intervals between breaks in the slope of the gage height graph, in minutes, and
$h_1, h_2, h_3, ..., h_n$		average gage height, in feet, during time interval 1, 2, 3, ..., n.

Using the data from figure 25, the computation of the time-weighted mean gage height is as follows:

Average gage height (h) in ft	Time interval (t) in minutes	$h \times t$
1.92	15	28.80
1.70	15	25.50
1.67	15	25.05
1.88	15	28.20
Totals	60	107.55

The mean gage height is computed as $H = 107.55/60 = 1.79$ ft.

In this example, there is little difference between the discharge-weighted mean gage height (1.77 ft) and the time-weighted mean gage height (1.79 ft). The average of the

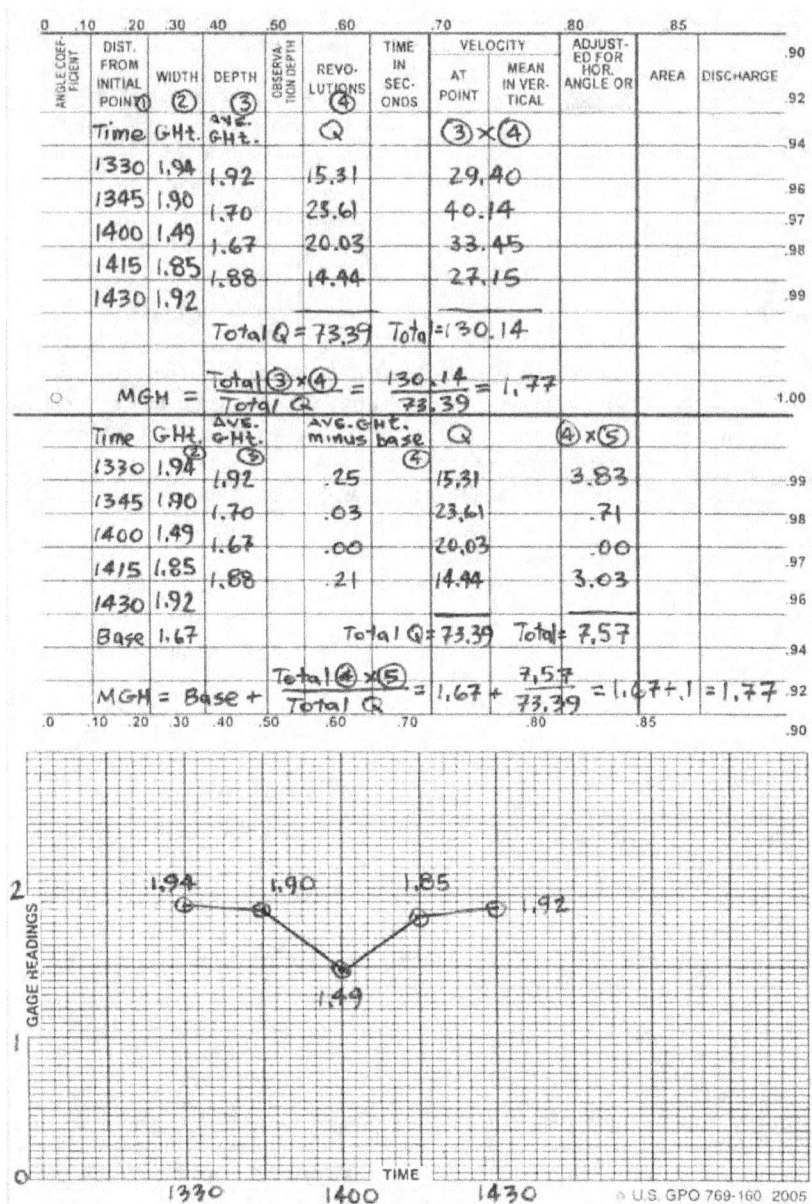

Figure 25.   Graph of readings used to compute a weighted mean gage height.

two values, 1.78 ft, is the preferred mean gage height for the discharge measurement.

When extremely rapid changes in stage occur during a measurement, the weighted mean gage height is not truly applicable to the discharge measured. To reduce the range in stage during the measurement, try to reduce the time required for making the discharge measurement; however, keep in mind that shortcuts in the measurement procedure usually reduce the accuracy of the measured discharge. Therefore, measurement procedures during rapidly changing stage must be optimized to produce a minimal combined error in measured discharge and computed mean gage height. The following section of this chapter describes procedures for making discharge measurements during periods of rapidly changing stage.

Discharge Measurements During Rapidly Changing Stage

Discharge measurements during periods of rapidly changing stage are more difficult to obtain, and accuracy is often not as good as for measurements made when the stage is fairly constant. The computation of total discharge and the corresponding stage are both subject to more uncertainty when the stage significantly changes during the period of the measurement. Two procedures are suggested for shortening the time required for a discharge measurement. The first procedure is applicable more for large streams where the stage changes are usually not as great as for small streams. The second procedure is designed more for the flash-flood conditions experienced with small streams, where peaks can be of momentary duration, and where the rising and falling stage is rapid. Keep in mind that the moving-boat ADCP discharge measurement can be used during these conditions. For a more detailed discussion of procedures for using an ADCP moving-boat during rapidly changing stage, see Mueller and Wagoner (2009).

Measurements of Large Streams With Rapidly Changing Stage

During periods of rapidly changing stage in a large stream, make measurements as quickly as possible to keep the change in stage to a minimum. This will minimize discharge errors caused by the shifting of flow patterns and other variables as the stage changes, and will provide a more accurate stage computation for the measurement. To reduce the time required for making a discharge measurement, make fewer than the usual number of observations and shorten the time to make them. This is sometimes referred to as a shortcut method. Following is a list of some of the things that can be done to reduce the time:

1. Use the 0.6-depth method rather than the 0.2 and 0.8 method. If the 0.6 method cannot be used because the flow is too swift, or if debris makes it too hazardous, then use the 0.2-depth method or the subsurface method. If an optical meter is available, use surface-velocity measurements.

2. Reduce the velocity observation time to about 20 to 30 seconds. This is referred to as half-counts.

3. Only measure about 15 to 18 sections. For some conditions, even use less than 15 sections may be used.

4. Observe and record the watch time at about every third vertical. If possible, observe and record the stage once or twice during the measurement.

By incorporating the above practices, a measurement can usually be made in 15 to 20 minutes. The surface or subsurface method for observing velocities is used, then some vertical-velocity curves will be needed later to establish coefficients to convert observed velocity to mean velocity. Compute a weighted mean gage height for the discharge measurement, as described in a previous section of this chapter.

Use the determination of error in discharge measurements, as described by Sauer and Meyer (1992), to illustrate the difference in errors between the standard measurement procedure and the shortcut procedure. The results shown in table 9 are based on a firm and smooth streambed. The additional uncertainty caused by using the shortcut method is generally less than the error that can be expected from the shifting of flow patterns and other variables that may occur during periods of rapidly changing stage.

Measurements of Small Streams With Rapidly Changing Stage

A series of "instantaneous" discharge measurements can be made during flash flood conditions on small streams by rating individual subsections, or verticals. This method requires repeated observations of gage height, depth, and velocity at selected verticals during the rise and fall of the flood wave. Two procedures are described below, with the primary difference being the method of determining depth at each vertical. In the first procedure, the streambed elevation referenced to gage datum is predetermined for each selected vertical. Depth is then determined as the difference between the gage height and the streambed elevation. In the second procedure, depth is measured at the selected verticals by sounding each time velocity is observed. The first procedure is faster; however, it may not be suitable if the streambed is unstable.

The method of computing the discharge measurements is also slightly different for the two procedures. For both procedures, a rating of gage height versus mean velocity is required

Table 9.   Comparison of discharge measurement error for standard and shortcut methods.

[No., number; s, seconds; ft/s, feet per second; ft, feet]

Measurement	Meter type	No. of verticals	Points in vertical	Suspension	Observation time, s	Mean velocity, ft/s	Mean depth, ft	Uncertainty percent	Measurement rated
Standard	AA	30	2	cable	50	2.50	10.0	2.3	Excellent/good.
Shortcut	AA	15	1	cable	25	2.50	10.0	5.1	Good/fair.

for each subsection, or vertical. For the second procedure, a rating of gage height versus depth is required for each vertical. The two procedures are described below.

Procedure 1—Depth Is Computed From Predetermined Streambed Elevations at Each Vertical

1. Select about 10 verticals, or subsections. For very small streams, you may use fewer verticals. Mark the selected verticals in some way so that repeated observations can be made at the same vertical each time.

2. Determine the streambed elevation referenced to gage datum for each selected vertical prior to making the series of discharge measurements. After the flood recedes, determine the streambed elevations at each vertical again to see if changes occurred during the flood. If the streambed is not stable, it will be necessary to interpolate the changes based on time and the best judgment of the hydrographer. Depth is determined at each vertical as the difference between this elevation and the gage height.

3. Take velocity observations at each vertical using the 0.6-depth method. Full counts of 40 seconds or more are recommended, but half-counts may be used if the stream is rising or falling extremely fast. If the 0.6 method cannot be used, then take velocity observations at the 0.2 depth or the subsurface depth. If an optical or other approve noncontact flowmeter is available, use it to take surface-velocity readings. For surface- and subsurface-velocity readings, it will be necessary to determine the coefficient required for converting the readings to a mean velocity. Meter positions should be based on the depth, as computed in item 2 above.

4. Make observations of other factors that would affect the computation of discharge, such as horizontal-angle coefficients.

5. Repeat the velocity and other observations at each of the selected verticals several times over the duration of the flood wave.

6. Record the watch time of each vertical measurement, and make corresponding gage- height observations frequently during the period of the flood wave.

7. Develop a rating of gage height versus mean velocity for each of the selected verticals. If surface- or subsurface-velocity observations were made, apply adjustments so that the rating represents mean velocity in the vertical. In some cases, it may be necessary to develop more than one rating for each vertical. For instance, a rating for the rising side of the flood wave, and a separate rating for the falling side of the flood wave, may be necessary.

8. Select a gage height for which a discharge measurement is to be computed. Use a standard discharge measurement note sheet for computing the discharge measurement.

Enter the stationing for the edge of water and for each of the selected verticals. Enter the depths at each vertical, computed on the basis of the selected gage height minus the streambed elevation. Enter the mean velocity at each vertical on the basis of the gage height versus mean velocity ratings. Enter other adjustments, such as horizontal-angle coefficients, as observed during the observation of velocities. Compute the discharge measurement similar to a regular discharge measurement.

9. Repeat the process described in item 8 above for other selected gage heights. If the ratings of gage height versus mean velocity change, such as for rising and falling stage, then compute separate discharge measurements for the rising and falling limbs of the flood wave.

Procedure 2—Depth Is Measured by Sounding at Each Vertical

1. Select about 10 verticals, or subsections, as described in the first procedure above. For very small streams, fewer verticals may be used. Mark the selected verticals in some way so that repeated observations can be made at the same vertical each time.

2. Determine the depth for each selected vertical by sounding the streambed each time you measure a vertical. Use this method when it is possible to easily make soundings, and when there is a likelihood of streambed elevation changes caused by scour or fill during the course of the measurement.

3. Take velocity observations at each vertical using the 0.6-depth method. Full counts of 40 seconds or more are recommended, but half-counts may be used if the stream is rising or falling extremely fast. It is not likely that you will need to make surface or subsurface observations because depth soundings are possible with this procedure. Meter positions should be based on the sounded depth.

4. Make observations of other factors that would affect the computation of discharge, such as horizontal-angle coefficients.

5. Repeat the observations of depth, velocity, and other variables at each of the selected verticals several times over the duration of the flood wave.

6. Record the watch time of each vertical measurement, and make frequent, corresponding gage-height observations during the period of the flood wave.

7. Develop a rating of gage height versus mean velocity for each of the selected verticals. As described in the first procedure, it may be necessary to develop more than one rating, such as for the rising and falling sides of the flood wave.

8. Develop a rating of gage height versus depth for each of the selected verticals. If streambed changes occur during the measurements, it will be necessary to take these into account by making appropriate corrections.

9. Select a gage height for which a discharge measurement is to be computed. Using a standard discharge-measurement note sheet, enter the stationing for the edge of water and for each of the selected verticals. Enter the depths at each vertical, based on the selected gage height and the ratings of gage height versus depth. Enter the mean velocity at each vertical on the basis of the gage height versus mean velocity ratings. Enter other adjustments, such as horizontal-angle coefficients, as observed during the observation of depths and velocities. Compute the discharge measurement similar to a regular discharge measurement.

10. Repeat the process described in item 9 above for other selected gage heights. If the depth and (or) mean velocity ratings change, such as for rising and falling stage, or for streambed scour or fill, then compute separate discharge measurements for conditions before and after the changes.

Correction of Discharge for Storage During Measurement

Most discharge measurements are made at or near the gaging station and the gage control. However, at some gages, it may be necessary to make discharge measurements at a substantial distance away from the gage and (or) control. For instance, during a flood, the only place to measure may be at a bridge located some distance from the gage. Or for some sites, the low-water section control may be located at a substantial distance downstream from the gage. If a discharge measurement is made at a substantial distance from the gage control during a change in stage, the discharge passing the control during the measurement will not be the same as the discharge at the measurement section. In these situations, an adjustment must be applied to the measured discharge to account for the change in channel storage that occurs between the measurement section and the control during the period of the measurement. The adjustment for channel storage is computed by multiplying the channel surface area by the average rate of change in stage in the reach between the measurement section and the control. The equation is

$$Q_G = Q_m \pm WL\frac{\Delta h}{\Delta t}, \qquad (13)$$

where Q_G discharge going over the control, in cubic feet per second,

Q_m measured discharge, in cubic feet per second,

W average width of stream between measuring section and control, in feet,

L length of reach between measuring section and control, in feet,

Δh average change in stage in the reach L during the measurement, in feet, and

Δt elapsed time during measurement, in seconds.

Determine the change in stage at each end of the reach (that is, at the control and at the measuring section) and use an average of these two values. Generally, the gage height at the gage is used at one end of the reach, and a reference point (RP) or a temporary gage is set at the other end of the reach. The water-surface elevation at each end of the reach is determined before and after the measurement to compute Δh. If the measurement is made upstream from the control, the adjustment will be plus for falling stages and minus for rising stages; if it is made downstream from the control, it will be minus for falling stages and plus for rising stages.

An example computation for a flood measurement that was made 0.6 mile upstream from the gage (and control) during a period of changing stage is shown below:

- Measurement made 0.6 mile upstream, L = 3,170 feet. Average width between measuring section and control, W = 150 ft.

- Gage height at beginning of measurement, at the gage (and control) = 5.84 ft.
  Gage height at end of measurement, at the gage (and control) = 6.74 ft.
  Change in stage at gage (and control), 6.74 − 5.84 = +0.90 ft.

- Gage height at beginning of measurement, at measuring section, = 12.72 ft.
  Gage height at end of measurement, at measuring section = 13.74 ft.
  Change in stage at measuring section, 13.74 − 12.72 = +1.02 ft.

Readings taken at measuring section from a reference point before and after measurement.

- Average change in stage in the reach, Dh = (0.90+1.02)/2 = 0.96 ft.
  Elapsed time during measurement, Dt = 1.25 hours = 4,500 seconds.
  Measured discharge, Q_m = 8,494 ft³/s.

$$Q_G = 8,494 - \left(150 \times 3,170 \times \frac{0.96}{4,500}\right) = 8,494 - 101 = 8,393\frac{ft^3}{s}, \quad (14)$$

This discharge should then be rounded to 8,390 ft³/s, which represents the discharge at the gage, or control.

The adjustment of the measured discharge for storage between the gage (or control) and measuring site, as described above, is a separate and distinct problem from that of making adjustments owing to variable water-surface slopes caused by changing discharge. Those adjustments are related to stage-discharge rating analysis, and are described by Kennedy (1984), and by Rantz (1982). The storage adjustment they described should be made immediately following the completion of the discharge measurement, and the resulting adjusted discharge is later used for rating analysis.

Instruments and Equipment

Point-velocity current-meter and profiler measurements are usually classified in terms of the method used to cross the stream during the measurement (that is, wading, bridge, cableway, or ice), the method used to compute the discharge, such as midsection, ADCP, flume, or volumetric, and the method used to compute the velocity (if applicable to the discharge method), such as Price AA meter, ADCP, or ADV. Instruments and equipment used in making current-meter measurements will vary, depending upon which of these measurement types are being used. Current meters, timers, and electronic and other counting equipment are generally common to all types of current-meter and profiler measurements. This section describes equipment currently used by USGS field offices.

Current Meters

A point-velocity current meter, in the context of this report, is a precision instrument calibrated to measure the velocity of flowing water in a single point or fixed volume. Several types of current meters are available for use, including rotating-element mechanical meters, electromagnetic meters, acoustic Doppler velocimeters (ADVs or FlowTrackers), acoustic digital current meters (ADCs), and optical meters.

The principle of operation for a mechanical meter is based on the proportionality between the velocity of the water and the resulting angular velocity of the meter rotor. By placing a mechanical current meter at a point in a stream and counting the number of revolutions of the rotor during a measured interval of time, the velocity of water at that point can be determined from the meter rating.

An electromagnetic current meter is based on the principle that a conductor (water) moving through a magnetic field will produce an electrical current. By measuring this current and the resultant distortion in the magnetic field, the instrument can be calibrated to determine point velocities of flowing water.

The acoustic meter or profiler (for example, ADV or ADCP) uses the Doppler principle to determine velocities of flowing water. Acoustic meters and profilers have been developed to measure point velocities in the vertical profile of an open-channel flow, as well as multicell vertical-velocity profiles. The ADCP has been adapted for use with the moving-boat method of measuring discharge, as described by Mueller and Wagner (2009). The ADV has been developed and adapted to mount on the standard USGS wading rod to measure point velocities in a manner similar to the method for measuring velocity with a vertical-axis rotating-element mechanical meter. ADVs will be described in a later section of this chapter.

The following sections describe the various types of current meters and profilers in more detail, give advantages and disadvantages of each, and provide guidance on care and maintenance.

Current Meters—Mechanical and Vertical Axis

Historically, the most commonly used meter by the USGS to measure open-channel velocities in rivers and streams has been the vertical-axis, mechanical current meter. The original prototype for this kind of current meter was designed and built in 1882 by W.G. Price, while he was working with the Mississippi River Commission. The Price current meter has evolved through a number of different models and refinements since 1882, but the basic theory and concepts remain the same. The Price AA meter is the most commonly used mechanical current meter for discharge measurements made by the USGS; however there are other variations of this meter, such as the Price AA slow velocity, the Price pygmy, and the Price AA winter meter. The following sections describe the various Price meters in more detail, and table 10 summarizes the various configurations and recommendations for the Price current meter, Price AA low velocity, the Price pygmy, the Price AA winter yoke with polymer and metal cups, and the SonTek FlowTracker ADV.

Price AA Meter

Historically, most current-meter measurements made by the USGS have been made with the vertical-axis Price AA and the Price pygmy current meters, as shown in figures 26 and 27. The basic components of the Price AA meter include the shaft and rotor (bucket wheel) assembly, the contact chamber, the yoke, and the tailpiece. The rotor, or bucket wheel, is 5 in. in diameter and 2 in. high with six cone-shaped cups mounted on a stainless-steel shaft. A vertical pivot supports the vertical shaft of the rotor, hence the name vertical-axis current meter. The contact chamber houses the upper part of the shaft and provides a method of counting the number of revolutions the rotor makes. A reduction gear (commonly referred to as the penta gear) on the lower part of the shaft allows counting every fifth revolution of the rotor when it is activated. The penta gear is used in discharge measurements with very high velocities. Contact chambers that can be used on the Price AA meter are described in a later section of this chapter. The yoke is the framework that holds the other components of the meter. A tailpiece is used for balance and keeps the meter pointing into the current.

When placed in flowing water, the rotors of the Price current meters turn at a speed proportional to the speed of the water. For practical purposes, these current meters are considered nondirectional because they register the maximum velocity of the water, even though they may be placed at an angle to the direction of flow. Advantages of the vertical-axis current meter are:

1. They operate in lower velocities than do horizontal-axis meters.

2. Bearings are well protected from silt-laden water.

3. The rotor is easily repairable in the field without adversely affecting the rating.

4. USGS standard ratings apply to the Price AA and Price pygmy meters.

5. A single rotor serves for the entire range of velocities.

Table 10. Price current meter and SonTek ADV configurations, usages, and recommended ranges of depth (without boundary interference) and velocity.

Meter	Contact chamber	Counting method	Rating	Velocity range, feet per second[1]	Depth range, feet	Remarks
Price AA	Standard, cat's whisker and penta gear	Headphones, CMD[2], or EFN[3]	Standard	0.2 to 12	1.25 or greater	The Price AA meter can be used as a low-velocity meter if equipped with an optic contact chamber.
			Individual	0.1 to 12		
	Magnetic	CMD[2] or EFN[3]	Standard	0.2 to 12		
			Individual	0.1 to 12		
	Optic		Standard or individual	0.1 to 12		
Price AA, low velocity	Cat's-whisker with double contact lobe on shaft. No penta gear.	Headphones, CMD[2], or EFN[3]	Individual	0.1 to 12	1.25 or greater	This is the traditional Price AA low velocity meter. An individual rating is recommended; however, a standard rating can be used if less accuracy is acceptable.
Price pygmy	Cat's whisker	Headphones, CMD[2], or EFN[3]	Standard or individual	0.2 to 4.0	0.3 to 1.5	--
Price, winter WSC[4] yoke, polymer cups	Cat's whisker	Headphones, CMD[2], or EFN[3]	Individual, with suspension device	0.1 to 12	1.25 or greater	This meter is recommended for conditions where slush ice is present.
	Magnetic or Optic	CMD[2] or EFN[3]				
Price, winter WSC[4] yoke, metal cups	Cat's whisker	Headphones, CMD[2], or EFN[3]	Individual, with suspension device	0.1 to 12	1.25 or greater	This meter is recommended for conditions where slush ice is not present.
	Magnetic or Optic	CMD[2] or EFN[3]				
SonTek FlowTracker ADV	N/A	N/A	Individual	0.003 to 13[5]	0.25 or greater	The FlowTracker has been documented by the USGS to provide velocities comparable to mechanical vertical axis current meters.

[1]Low- and high-velocity limits shown in the table are based on a small-to-moderate extrapolation of the lower and upper meter calibration limits. It is not recommended that the meters be used for velocities less than the lower limit. The velocity rating for the Price meter may allow additional extrapolation in the upper range to about 20 feet per second. The upper range of the Price pygmy meter rating may be extrapolated to about 5 feet per second. Standard errors within the meter calibration limits are less than ±5% in all cases. Standard errors in the extrapolated range of velocities are unknown, but are probably within ± 5%.

[2]Current-meter digitizer. Observe cautions for low velocities. See text.

[3]Electronic field notebook, such as Aquacalc or DMX. Observe cautions for low velocities. See text.

[4]Water Survey of Canada.

[5]Manufacturer's specification.

Price AA Meter (Slow Velocity)

In addition to the Price AA meter described above, there is a Price AA meter modified slightly for use in measuring low velocities. To reduce friction, the penta gear has been removed from this meter, and the shaft has two eccentrics making two contacts with the cat's whisker per revolution. The low-velocity meter normally is rated from 0.2 to 2.5 ft/s and is recommended when the mean velocity at a cross section is less than 1 ft/s.

Price Pygmy Meter

A miniature version of the Price AA meter is the Price pygmy meter, as shown in figures 26 and 27, which is used for measuring velocities in shallow depths. The Price pygmy meter is scaled two-fifths the size of the standard meter and has neither a tailpiece nor a penta gear. The contact chamber is an integral part of the yoke of the meter. The Price pygmy meter makes one contact for each revolution and is used only for rod suspension.

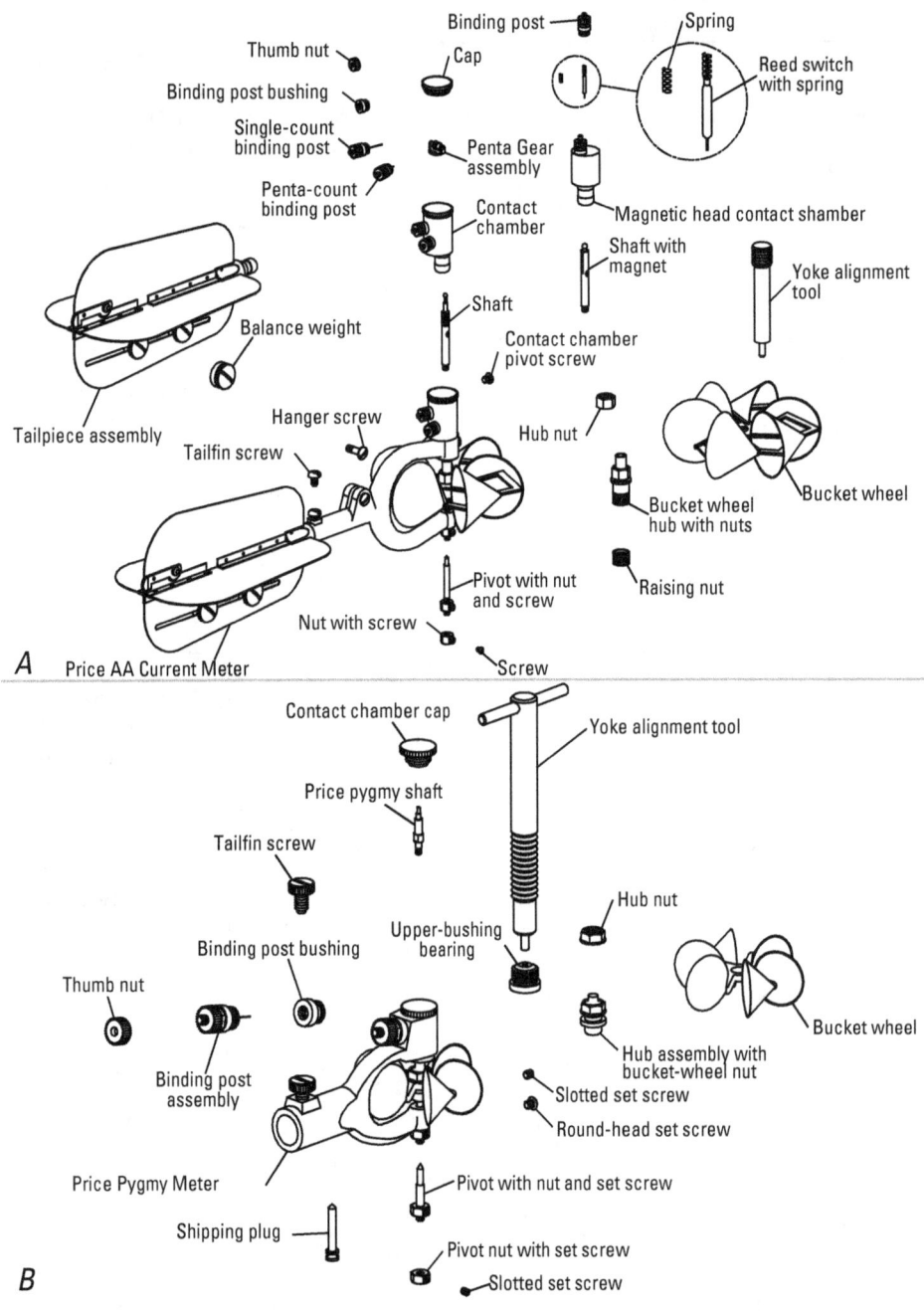

Figure 26. Assembly drawing of the *A*, Price AA current meter and *B*, Price pygmy current meter.

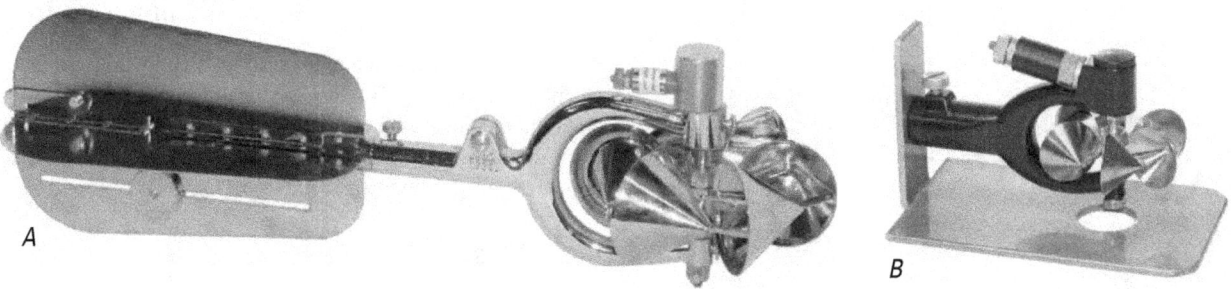

Figure 27. *A*, Price AA and *B*, Price pygmy current meters.

Price AA Winter Meter

In streams where slush ice is present, use a modified Price AA meter, as shown in figure 28. This meter is built with a WSCan winter-style yoke, and uses a polymer rotor (bucket wheel) in place of the standard metal-cup rotor. The WSCan winter-style yoke meters require an individual rating due to increased tolerances in manufacturing. The solid polymer rotor has the advantage that it does not fill with slush ice during a measurement, and the slush ice does not easily adhere to it. If slush ice is not present, alternative methods can be used, including replacing the polymer rotors with metal cups or using a pygmy meter. Regular Price AA meters with metal-cup rotors are also acceptable for slush-free conditions, if there are no problems with cutting the required larger holes through the ice. In recent years, the USGS has increasingly used ADCPs and ADVs as alternatives to the Price AA for making discharge measurements in streams where ice is present.

Figure 28. Price AA meter with winter-style yoke and polymer rotor.

Current Meters—Mechanical and Horizontal Axis

A number of mechanical current meters are available that have a propeller-, or vane-, type of rotor mounted on a horizontal shaft. These meters are used extensively in Europe and some Asian countries, but very little in the United States, and are generally not recommended by the USGS because they are not as durable as the Price current meter. Horizontal-axis current meters include the Ott (Germany), Neyrpic (France), Haskell (U.S.), Hoff (U.S.), Braystoke (United Kingdom), and Valeport. Various models of each of these are also available. As a group, horizontal-axis current meters have the following advantages:

1. The rotor, or propeller, disturbs flow less than vertical-axis rotors because of axial symmetry with flow direction.

2. The rotor is less likely to be entangled by debris than vertical-axis rotors.

3. Bearing friction is less than for vertical-axis rotors because bending moments on the rotor are eliminated.

4. In oblique currents, some of these meters (for example, the Ott meter) measure the velocity normal to the measuring section when the meter is held normal to the measuring section.

5. Rotors with propellers of different pitches are available for some of the meters, allowing measurement of a considerable range of velocity.

See figures 29, 30, and 31, respectively, for examples of the Ott, Hoff, and Valeport current meters.

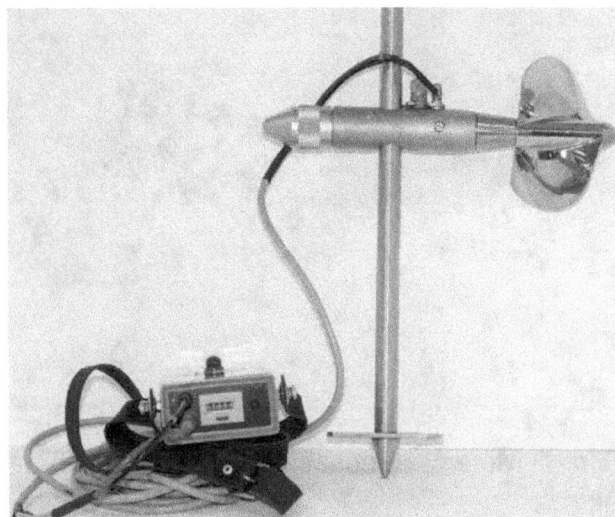

Figure 29.   Ott current meter.

Figure 30.   Hoff current meter.

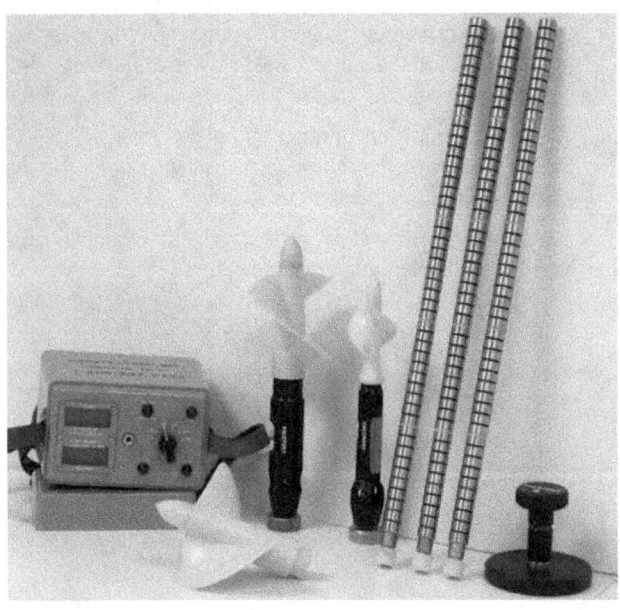

Figure 31.   Valeport current meter.

Current-Meter Contact Heads (Vertical Axis)

The Price current meter is normally fitted with a contact chamber that has a cat's-whisker type of circuitry used for counting the number of revolutions of the rotor. Two other types of contact chambers—the magnetic switch type and the optical type—can be fitted to the Price AA meter.

Cat's Whisker

When placed in flowing water, the rotor of the current meter turns at a speed proportional to the speed of the water. The number of revolutions of the rotor is obtained by counting electrical impulses generated in the contact chamber. An eccentric contact on the upper end of the rotor shaft wipes a slender alloy wire (cat's whisker) attached to the binding post, which closes an electrical circuit. This electrical impulse produces an audible click in a headphone or registers a unit on a counting device. Contact points in the chamber are designed to complete the electrical circuit at selected frequencies of revolution, such as twice per revolution, once per revolution, or once per five revolutions (by use of the penta gear). The penta gear, wire, and binding post provide a contact each time the rotor makes five revolutions. Figure 32 shows the contact chamber and shaft for the cat's-whisker-type chamber, with a single-count binding post.

Two types of cat's-whisker wires have been used: one is the simple bronze wire, and the other is the old type of wire with a small solder bead on the end of it. Adjust the cat's whisker for the penta gear so that it always touches the penta eccentric, even when the penta counter is not in use. Otherwise, the meter rating may be affected.

Magnetic Switch

A contact chamber housing a magnetic-type switch, as shown in figure 32, is available to replace the cat's-whisker-contact chamber. The magnetic switch is composed of glass that is enclosed in a hydrogen atmosphere and hermetically sealed. The switch assembly is rigidly fixed in the top of the meter head just above the tip of the shaft. The switch is operated by a small permanent magnet rigidly fastened to the shaft. Two types of magnets used: (1) a bar magnet and (2) a circular magnet. If the contact chamber uses the bar magnet, it should be identified with an "A" stamped on the top surface of the chamber to indicate it has been modified. Older, unmodified contact chambers with the bar magnet were found to under-register for velocities greater than about 2 ft/s. The chambers that utilize the circular magnets fit the standard rating throughout its range.

The magnetic switch quickly closes when the magnet is aligned with it, and then promptly opens when the magnet moves away. The magnet is properly balanced on the shaft. Any type of AA meter can have a magnetic switch added by replacing the shaft and the contact chamber. The magnetic switch is placed in the contact chamber through the tapped hole for the binding post. The rating of the meter is not altered by the change.

An automatic counter, as described in a later section of this chapter, is used with the magnetic-switch contact chamber. Do not use a headphone with the magnetic head because arcing can weld the contacts.

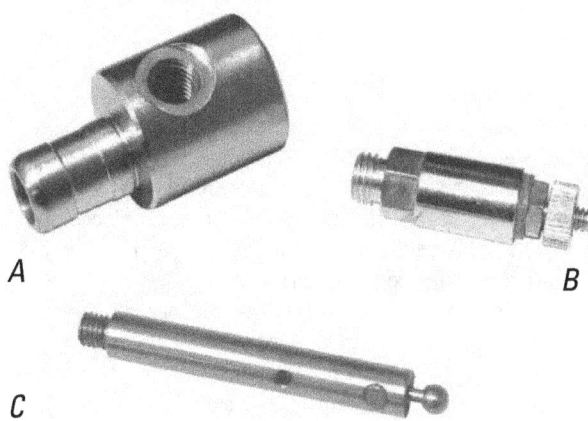

Figure 32.   *A*, Price AA magnetic head current meter contact chamber; *B*, single count binding post; and *C*, shaft with magnet for magnetic heads.

Optical Head

A contact head utilizing fiber-optics technology is available for reading the pulse rate of the Price AA current meter. A special rotor containing two fiber-optic bundles is attached to the upper end of the bucket-wheel shaft. The rotation of these fiber-optic bundles gates infrared light from a photo diode to a photo transistor, creating a pulse rate that is proportional to the rotor's revolutions. The pulses are counted, stored, and then compared with a quartz crystal oscillator. This information is processed to display stream velocity on a liquid crystal readout. The display has three averaging periods selected by a rotary switch. The averaging periods range from a minimum of about 5 seconds to a maximum of about 90 seconds. The unit is powered by a 9-volt battery.

Output of pulses from the optical sensing unit can be counted by the current-meter digitizer and the electronic field notebooks described in subsequent sections of this chapter. A standard rating table based on tow-tank calibration tests at the USGS Hydrologic Instrumentation Facility (HIF) is used to convert pulse rate to stream velocity.

A special tail-fin assembly is required for the optical meter so it will balance properly when submerged. The vertical section of this tail fin is marked with the letters OAA, and the horizontal section is marked PAA.

Vertical-Axis Current-Meter Timers and Counters

The determination of velocity using a mechanical current meter requires that the number of revolutions of the rotor be counted during a specified time interval, usually 40 to 70 seconds. Several methods are available for timing and counting the revolutions, as described in the following paragraphs.

Stopwatch and Headset

For current meters having a cat's-whisker-type contact chamber, an electrical circuit is closed each time the contact wire touches the single or penta eccentric of the current meter. A battery and headphone, as shown in figure 33, are parts of the electrical circuit, and an audible click can be heard in the headphone at each electrical closure. Some hydrographers have adapted compact, comfortable hearing-aid-type phones to replace headphones. Beepers that can be heard without the headset are also sometimes used. Do not use a headset, or similar device, with the magnetic contact chamber because arcing can weld the contacts.

Measure the time interval to the nearest second with a stopwatch. Figure 33 shows the standard analog stopwatch; however, a digital wrist watch can also be used.

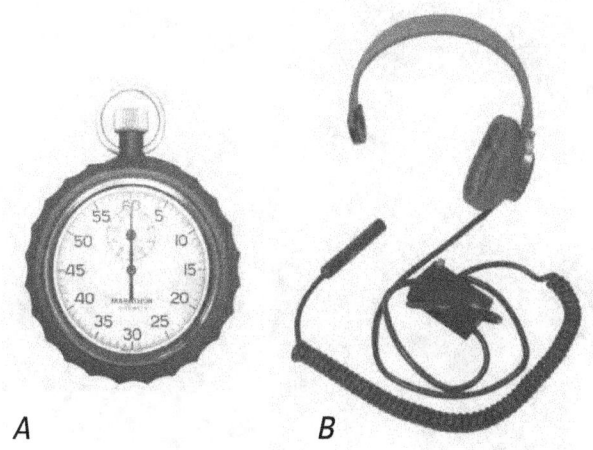

Figure 33.   *A*, Analog stopwatch and *B*, current-meter headset.

Electronic Counters

Current-Meter Digitizer (CMD)

A current-meter digitizer (CMD), or automatic electronic counter, as shown in figure 34, was developed for use with the cat's-whisker-, optic-, and magnetic-contact chambers. It can be used with any of the mechanical, vertical-axis current meters, but be careful to avoid false counts when using it for low velocities when the cat's-whisker-contact chamber is used. The CMD automatically counts and displays the number of revolutions of the current-meter rotor and the elapsed time. A buzzer produces an audible signal at each contact closure, and the total counts and elapsed time are shown in the display at the completion of the velocity measurement. A coded chip can be installed in the CMD that will compute and display the velocity from the standard rating table for the particular meter being used. The CMD is powered with five rechargeable batteries, and has an adapter that can be used to attach it to the top of a wading rod.

Figure 34.   Current-meter digitizer.

Other Electronic Counters, Electronic Field Notebooks, and Personal Digital Assistants

Electronic counters and timers for mechanical current meters are also available in some commercially available devices. Electronic field notebooks (EFNs), such as the JBS Instruments Aquacalc Pro Discharge Measurement Computer, are designed for electronic recording of discharge-measurement data. These will be described in more detail in a later section of this chapter. Both the Aquacalc and the Hydrological Services Current Meter Counter signal processor (CMCsp) contain built-in digitizers that count and time the current-meter rotor revolutions. Built-in ratings convert the revolutions and elapsed time to velocity. Just as with the CMD described above, the EFNs should be used with caution to avoid false counts when measuring low velocities with meters equipped with a cat's-whisker-contact chamber.

The USGS is increasing development and use of personal digital assistants (PDAs) for primary data collection and processing of discharge measurements and surface-water field data. Software programs such as the Surface WAter Measurement and Inspection application (SWAMI) continue to streamline field activities by reducing the need for paper forms in the field through the development of electronic field forms written for the PDA. SWAMI, as described in previous sections of this chapter, is used in many USGS Water Science Centers, and other PDA applications and programs are continually being developed.

Care of the Vertical-Axis Mechanical Current Meter

There are a number of documents describing the care and maintenance of the vertical-axis current meters. Among these, the most important originated from the USGS and are by Smoot and Novak (1968) and by Rantz (1982), and Office of Surface Water Technical Memorandum No. 89.07 (1989) and Office of Surface Water Technical Memorandum No. 99.06 (1999). These instructions represent a long history of experience based on field use of the meters, as well as from individuals in the Office of Surface Water Hydraulic Laboratory who have repaired and adjusted current meters to calibrate them within close tolerances. A brief description of the recommended procedure for checking the condition of a current meter, and for its care and cleaning during daily field use, is presented in the next few paragraphs. For complete details, consult the above-mentioned documents.

Recommended Procedure Before, During, and After Each Discharge Measurement

1. Before each discharge measurement, make a visual examination of the meter cups or vanes, pivot, bearing, and shaft for damage, wear, or faulty alignment. Inspect the bearing surface for water. This will usually appear as a milky emulsification of oil and water on the lower bearing and pivot, and in the contact chamber. If water is found, dry the meter parts and re-oil because the presence of water will affect the performance of the meter. The lower bearing is probably the most susceptible to the entrance of water.

2. Spin the rotor to make certain it operates freely, and allow it to slowly return to a resting position. If the rotor does not turn smoothly, or if it stops abruptly, then it is a sign of some problem and it should be corrected before using the meter. Check the balance and alignment of the meter on the hanger or wading rod. Be sure that the conductor wire does not interfere with meter balance and rotor spin.

3. During measurements, check the meter periodically when it is out of the water to be sure that the rotor spins freely, and that there is no debris or other substance obstructing it.

4. After a measurement is completed, make another visual inspection as described above to ensure that nothing was damaged or caused the meter to malfunction during the measurement. If there is a problem, you may have to make another discharge measurement.

5. Timed spin tests (described later in this chapter) are not required for each discharge measurement. The visual inspection described above is preferred over timed spin tests made in the field.

Recommended Procedure After a Day of Use in the Field

1. Examine the pivot and bearing surfaces for wear and damage, especially the pivot point. The pivot should feel sharp, not rounded or dull. It should not have a burr detectable visually or with the fingernail. A magnifying glass is helpful in making this examination. If the pivot is dull or burred, replace it with a new one.

2. Clean and lightly oil the pivot, bearing, and upper shaft with current-meter oil. Do not use regular machine oil, such as "3-in-1," because it tends to become gummy when exposed to water.

3. Check and carefully adjust cat's-whisker contacts, if necessary. Cat's whiskers should be made of simple bronze wire, not beaded wire.

4. After replacing the contact-chamber cap, spin the meter to see if it is operating correctly, as previously described. A timed spin test is not required.

Recommended Procedure After Each Field Trip

After each field trip, completely disassemble, inspect, and clean current meters. Make any necessary repairs. Detailed instructions for the disassembly, inspection, and adjustment of Price AA (both standard and magnetic head) and pygmy current meters are contained in the attachments to Office of Surface Water Technical Memorandum No. 99.06 (1999). A timed spin test may also be performed after each field trip, and after meter repairs.

Inactive Current Meters

Disassemble, inspect, and clean current meters as described above, prior to storing them. If the period of storage is less than 1 year, the meter may be used without further maintenance if an inspection and a spin test indicate it is operating properly. If the meter has been in storage longer than 1 year, or an indeterminate period, complete inspection, cleaning, and adjustment before using the meter.

Spin Tests

A timed spin test, made in the field before and after each discharge measurement, is no longer a requirement as it was in the past. The visual tests as described above are adequate for checking the meter in the field. Note "OK" or "free" in the spaces on the front sheet of the discharge measurement for spin test information to indicate that the visual check of the meter was acceptable.

Perform the full-timed spin test under controlled conditions between field trips, when the meter is suspect, and before and after repairs. Place the meter on a stable, level surface to perform the spin test. There should be no wind currents or drafts that can affect the rotor spin. Sharply spin the rotor while starting a stopwatch. Stop the stopwatch when the rotor comes to a complete stop. The minimum, acceptable spin times are as follows:

All types of Price pygmy meters	0:45 seconds
All types of Price AA meters	2:00 minutes

These are considered to be absolute minimum spin times. Meters in good condition will perform substantially better.

Recordkeeping

Maintain a current-meter log to record the results of the timed spin tests for each current meter. In addition, the log should contain information that identifies the meter and rotor, a history of repairs to the meter, as well as the name of the person who checked the meter, and dates of occurrences. Figure 35 shows a recommended format for the current meter log. The current-meter log should become a permanent record and archived with other water-resources data.

CURRENT METER LOG				
Meter Type: AA Pygmy Other_____ (Circle one)			Meter No._____	Rotor No._____
Date	Meter User	Entry made by	Spin Time	Description of repairs, notation of disassembly, inspections, and remarks
			:	
			:	
			:	
			:	
			:	
			:	
			:	
			:	
			:	
			:	
			:	
			:	
			:	
			:	
			:	
			:	
			:	
			:	
			:	
			:	
			:	
			:	
			:	
			:	
			:	
			:	
			:	
			:	
			:	
			:	
			:	
			:	
			:	
			:	
			:	
			:	
			:	
			:	
			:	
			:	
			:	
			:	
			:	

Figure 35. Example of a current-meter log (suggested format).

Rating of Mechanical Current Meters

In order to determine the velocity of the water from the revolutions of the rotor of a mechanical current meter, a relation must be established between the angular velocity of the rotor and the velocity of the water turning it. This relation is referred to as the current-meter rating, and is expressed in an equation or in tabular format.

The current-meter rating facility is operated by the USGS, and is located at the Hydraulics Laboratory of the HIF at Stennis Space Center, Miss. The rating facility consists of a sheltered, reinforced concrete basin 400 ft long, 6 ft wide, and 6 ft deep, commonly called the tow tank. An electrically driven car rides on rails alongside and extending the length of the water-filled basin, and carries the current meter at a constant rate through the still water. Although the rate of travel can be accurately adjusted by means of a hydraulic regulating gear, the average velocity of the moving car is determined for each run by making an independent measurement of the distance it travels during the time that the revolutions of the rotor are electrically counted. A scale graduated in feet and tenths of a foot is used for this purpose. Eight pairs of runs are usually made for each current meter. A pair of runs consists of two traverses of the basin, one in each direction, at approximately the same speed. Practical considerations usually limit the ratings to velocities ranging from 0.1 ft/s to about 15 ft/s, although the rating car can be operated at lower speeds. Unless a special request is made for a more extensive rating, the lowest velocity used in the rating is about 0.2 ft/s, and the highest is about 8.0 ft/s.

Because there is rigid control in the manufacture of the Price meter, virtually identical meters are produced and, for practical purposes, their rating equations are identical. Therefore, there is no need to calibrate the meters individually, which is a major advantage and time saver. Instead, a standard rating is established by calibrating a group of meters that have been constructed according to USGS specifications. This standard rating is essentially an average rating for the calibration group, and it is then supplied with all meters manufactured according to USGS specifications. Identicalness of meters is ensured by supplying the dies and fixtures for the construction of Price current meters to the manufacturer who makes the successful bid. Another advantage of the standard rating is that field repairs can be made to a meter without requiring that it be recalibrated. On the other hand, there are somewhat larger errors associated with the standard ratings, as opposed to the individual meter ratings. For additional details see Office of Surface Water Technical Memoranda Nos. 91.01 (1991) and 99.05 (1999).

Standard current-meter ratings are not mandatory for use with the Price meters. For some applications, it may be appropriate to use individually rated meters to avoid the additional uncertainty of the standard ratings. All winter-style meters must be individually rated with the suspension device that will be used with it.

Standard current-meter ratings, as of 1999, have been defined for the Price AA with the cat's-whisker- and magnetic-contact chambers, and the Price pygmy with the cat's-whisker-contact chamber. The standard rating for the Price AA with the fiber-optic-contact chamber was defined in 1991. These ratings are as follows:

- Price AA with cat's-whisker- and magnetic-contact chambers (Standard rating No. 2)

$$V = 2.2048R + 0.0178 \qquad (15)$$

- Price pygmy with cat's-whisker-contact chamber (Standard rating No. 2)

$$V = 0.9604R + 0.0312 \qquad (16)$$

- Price AA with fiber-optic-contact chamber

$$V = 2.194R + 0.014 \ (R<0.856) \qquad (17)$$

$$V = 2.162R + 0.041 \ (R>0.856) \qquad (18)$$

where V velocity, in feet per second (ft/s), and
R the number of rotor revolutions per second.

For convenience in field use, the data from the current-meter ratings are reproduced in tables, samples of which are shown in figures 36 and 37. In figure 36, the velocities corresponding to a range of 3 to 200 revolutions of the rotor within a period of 40 to 70 seconds are listed in the tables.

In figure 37, the velocities corresponding to a range of 3 to 350 revolutions of the rotor within a period of 40 to 70 seconds are listed in the tables. This range in revolution and time has been found to cover general field requirements. To provide the necessary information for extending a table for the few instances where extensions are required, the equation of the rating table is shown in the heading.

Meters that have been rated by means of rod suspension, and then by means of cable suspension using Columbus-type weights and hangers, have not shown significant differences in their ratings. Therefore, no suspension coefficients are needed if weights and hangers are properly used.

The preceding discussion relates primarily to the Price current meters. Other types of meters, such as the horizontal-axis meters and the electromagnetic meters, can also be calibrated in the tow tank in a similar manner as the Price meters. The HIF designed and constructed a special tow tank for testing and calibrating the acoustic Doppler point-velocity FlowTracker velocimeter. In the near future, there are HIF plans to build and operate a tow tank designed to test and calibrate ADCPs.

STANDARD RATING TABLE NO. 2 FOR PYGMY CURRENT METER (6/99)
EQUATION: V = 0.9604 R+ 0.0312 (R=revolutions per second)

Seconds	VELOCITY IN FEET PER SECOND Revolutions														
	3	5	7	10	15	20	25	30	40	50	60	80	100	150	200
40	0.103	0.151	0.199	0.271	0.391	0.511	0.631	0.752	0.992	1.23	1.47	1.95	2.43	3.63	4.83
41	0.101	0.148	0.195	0.265	0.383	0.500	0.617	0.734	0.968	1.20	1.44	1.91	2.37	3.54	4.72
42	0.100	0.146	0.191	0.260	0.374	0.489	0.603	0.717	0.946	1.17	1.40	1.86	2.32	3.46	4.60
43	0.098	0.143	0.188	0.255	0.366	0.478	0.590	0.701	0.925	1.15	1.37	1.82	2.26	3.38	4.50
44	0.097	0.140	0.184	0.249	0.359	0.468	0.577	0.686	0.904	1.12	1.34	1.78	2.21	3.31	4.40
45	0.095	0.138	0.181	0.245	0.351	0.458	0.565	0.671	0.885	1.10	1.31	1.74	2.17	3.23	4.30
46	0.094	0.136	0.177	0.240	0.344	0.449	0.553	0.658	0.866	1.08	1.28	1.70	2.12	3.16	4.21
47	0.093	0.133	0.174	0.236	0.338	0.440	0.542	0.644	0.849	1.05	1.26	1.67	2.07	3.10	4.12
48	0.091	0.131	0.171	0.231	0.331	0.431	0.531	0.631	0.832	1.03	1.23	1.63	2.03	3.03	4.03
49	0.090	0.129	0.168	0.227	0.325	0.423	0.521	0.619	0.815	1.01	1.21	1.60	1.99	2.97	3.95
50	0.089	0.127	0.166	0.223	0.319	0.415	0.511	0.607	0.800	0.992	1.18	1.57	1.95	2.91	3.87
51	0.088	0.125	0.163	0.220	0.314	0.408	0.502	0.596	0.784	0.973	1.16	1.54	1.91	2.86	3.80
52	0.087	0.124	0.160	0.216	0.308	0.401	0.493	0.585	0.770	0.955	1.14	1.51	1.88	2.80	3.73
53	0.086	0.122	0.158	0.212	0.303	0.394	0.484	0.575	0.756	0.937	1.12	1.48	1.84	2.75	3.66
54	0.085	0.120	0.156	0.209	0.298	0.387	0.476	0.565	0.743	0.920	1.10	1.45	1.81	2.70	3.59
55	0.084	0.119	0.153	0.206	0.293	0.380	0.468	0.555	0.730	0.904	1.08	1.43	1.78	2.65	3.52
56	0.083	0.117	0.151	0.203	0.288	0.374	0.460	0.546	0.717	0.889	1.06	1.40	1.75	2.60	3.46
57	0.082	0.115	0.149	0.200	0.284	0.368	0.452	0.537	0.705	0.874	1.04	1.38	1.72	2.56	3.40
58	0.081	0.114	0.147	0.197	0.280	0.362	0.445	0.528	0.694	0.859	1.02	1.36	1.69	2.51	3.34
59	0.080	0.113	0.145	0.194	0.275	0.357	0.438	0.520	0.682	0.845	1.01	1.33	1.66	2.47	3.29
60	0.079	0.111	0.143	0.191	0.271	0.351	0.431	0.511	0.671	0.832	0.992	1.31	1.63	2.43	3.23
61	0.078	0.110	0.141	0.189	0.267	0.346	0.425	0.504	0.661	0.818	0.976	1.29	1.61	2.39	3.18
62	0.078	0.109	0.140	0.186	0.264	0.341	0.418	0.496	0.651	0.806	0.961	1.27	1.58	2.35	3.13
63	0.077	0.107	0.138	0.184	0.260	0.336	0.412	0.489	0.641	0.793	0.946	1.25	1.56	2.32	3.08
64	0.076	0.106	0.136	0.181	0.256	0.331	0.406	0.481	0.631	0.782	0.932	1.23	1.53	2.28	3.03
65	0.076	0.105	0.135	0.179	0.253	0.327	0.401	0.474	0.622	0.770	0.918	1.21	1.51	2.25	2.99
66	0.075	0.104	0.133	0.177	0.249	0.322	0.395	0.468	0.613	0.759	0.904	1.20	1.49	2.21	2.94
67	0.074	0.103	0.132	0.175	0.246	0.318	0.390	0.461	0.605	0.748	0.891	1.18	1.46	2.18	2.90
68	0.074	0.102	0.130	0.172	0.243	0.314	0.384	0.455	0.596	0.737	0.879	1.16	1.44	2.15	2.86
69	0.073	0.101	0.129	0.170	0.240	0.310	0.379	0.449	0.588	0.727	0.866	1.14	1.42	2.12	2.81
70	0.072	0.100	0.127	0.168	0.237	0.306	0.374	0.443	0.580	0.717	0.854	1.13	1.40	2.09	2.78
	3	5	7	10	15	20	25	30	40	50	60	80	100	150	200

Figure 36. Example of a standard current-meter rating table No. 2 for Price pygmy current meters with cat's-whisker contact chamber.

STANDARD RATING TABLE NO. 2 FOR AA CURRENT METERS (6/99)
EQUATION: V = 2.2048 R + 0.0178 (R=revolutions per second)

VELOCITY IN FEET PER SECOND

Seconds	50	60	80	100	150	200	250	300	350	Seconds
	Revolutions									
40	2.77	3.33	4.43	5.53	8.29	11.04	13.80	16.55	19.31	40
41	2.71	3.24	4.32	5.40	8.08	10.77	13.46	16.15	18.84	41
42	2.64	3.17	4.22	5.27	7.89	10.52	13.14	15.77	18.39	42
43	2.58	3.09	4.12	5.15	7.71	10.27	12.84	15.40	17.96	43
44	2.52	3.02	4.03	5.03	7.53	10.04	12.55	15.05	17.56	44
45	2.47	2.96	3.94	4.92	7.37	9.82	12.27	14.72	17.17	45
46	2.41	2.89	3.85	4.81	7.21	9.60	12.00	14.40	16.79	46
47	2.36	2.83	3.77	4.71	7.05	9.40	11.75	14.09	16.44	47
48	2.31	2.77	3.69	4.61	6.91	9.20	11.50	13.80	16.09	48
49	2.27	2.72	3.62	4.52	6.77	9.02	11.27	13.52	15.77	49
50	2.22	2.66	3.55	4.43	6.63	8.84	11.04	13.25	15.45	50
51	2.18	2.61	3.48	4.34	6.50	8.66	10.83	12.99	15.15	51
52	2.14	2.56	3.41	4.26	6.38	8.50	10.62	12.74	14.86	52
53	2.10	2.51	3.35	4.18	6.26	8.34	10.42	12.50	14.58	53
54	2.06	2.47	3.28	4.10	6.14	8.18	10.23	12.27	14.31	54
55	2.02	2.42	3.22	4.03	6.03	8.04	10.04	12.04	14.05	55
56	1.99	2.38	3.17	3.95	5.92	7.89	9.86	11.83	13.80	56
57	1.95	2.34	3.11	3.89	5.82	7.75	9.69	11.62	13.56	57
58	1.92	2.30	3.06	3.82	5.72	7.62	9.52	11.42	13.32	58
59	1.89	2.26	3.01	3.75	5.62	7.49	9.36	11.23	13.10	59
60	1.86	2.22	2.96	3.69	5.53	7.37	9.20	11.04	12.88	60
61	1.83	2.19	2.91	3.63	5.44	7.25	9.05	10.86	12.67	61
62	1.80	2.15	2.86	3.57	5.35	7.13	8.91	10.69	12.46	62
63	1.77	2.12	2.82	3.52	5.27	7.02	8.77	10.52	12.27	63
64	1.74	2.08	2.77	3.46	5.19	6.91	8.63	10.35	12.08	64
65	1.71	2.05	2.73	3.41	5.11	6.80	8.50	10.19	11.89	65
66	1.69	2.02	2.69	3.36	5.03	6.70	8.37	10.04	11.71	66
67	1.66	1.99	2.65	3.31	4.95	6.60	8.24	9.89	11.54	67
68	1.64	1.96	2.61	3.26	4.88	6.50	8.12	9.74	11.37	68
69	1.62	1.94	2.57	3.21	4.81	6.41	8.01	9.60	11.20	69
70	1.59	1.91	2.54	3.17	4.74	6.32	7.89	9.47	11.04	70
	50	60	80	100	150	200	250	300	350	

STANDARD RATING TABLE NO. 2 FOR AA CURRENT METERS (6/99)
EQUATION: V = 2.2048 R + 0.0178 (R=revolutions per second)

VELOCITY IN FEET PER SECOND

Seconds	3	5	7	10	15	20	25	30	40	Seconds
	Revolutions									
40	0.183	0.293	0.404	0.569	0.845	1.12	1.40	1.67	2.22	40
41	0.179	0.287	0.394	0.556	0.824	1.09	1.36	1.63	2.17	41
42	0.175	0.280	0.385	0.543	0.805	1.07	1.33	1.59	2.12	42
43	0.172	0.274	0.377	0.531	0.787	1.04	1.30	1.56	2.07	43
44	0.168	0.268	0.369	0.519	0.769	1.02	1.27	1.52	2.02	44
45	0.165	0.263	0.361	0.508	0.753	0.998	1.24	1.49	1.98	45
46	0.162	0.257	0.353	0.497	0.737	0.976	1.22	1.46	1.94	46
47	0.159	0.252	0.346	0.487	0.721	0.956	1.19	1.43	1.89	47
48	0.156	0.247	0.339	0.477	0.707	0.936	1.17	1.40	1.86	48
49	0.153	0.243	0.333	0.468	0.693	0.918	1.14	1.37	1.82	49
50	0.150	0.238	0.326	0.459	0.679	0.900	1.12	1.34	1.78	50
51	0.147	0.234	0.320	0.450	0.666	0.882	1.10	1.31	1.75	51
52	0.145	0.230	0.315	0.442	0.654	0.866	1.08	1.29	1.71	52
53	0.143	0.226	0.309	0.434	0.642	0.850	1.06	1.27	1.68	53
54	0.140	0.222	0.304	0.426	0.630	0.834	1.04	1.24	1.65	54
55	0.138	0.218	0.298	0.419	0.619	0.820	1.02	1.22	1.62	55
56	0.136	0.215	0.293	0.412	0.608	0.805	1.00	1.20	1.59	56
57	0.134	0.211	0.289	0.405	0.598	0.791	0.985	1.18	1.57	57
58	0.132	0.208	0.284	0.398	0.588	0.778	0.968	1.16	1.54	58
59	0.130	0.205	0.279	0.391	0.578	0.765	0.952	1.14	1.51	59
60	0.128	0.202	0.275	0.385	0.569	0.753	0.936	1.12	1.49	60
61	0.126	0.199	0.271	0.379	0.560	0.741	0.921	1.10	1.46	61
62	0.124	0.196	0.267	0.373	0.551	0.729	0.907	1.08	1.44	62
63	0.123	0.193	0.263	0.368	0.543	0.718	0.893	1.07	1.42	63
64	0.121	0.190	0.259	0.362	0.535	0.707	0.879	1.05	1.40	64
65	0.120	0.187	0.255	0.357	0.527	0.696	0.866	1.04	1.37	65
66	0.118	0.185	0.252	0.352	0.519	0.686	0.853	1.02	1.35	66
67	0.117	0.182	0.248	0.347	0.511	0.676	0.840	1.01	1.33	67
68	0.115	0.180	0.245	0.342	0.504	0.666	0.828	0.991	1.31	68
69	0.114	0.178	0.241	0.337	0.497	0.657	0.817	0.976	1.30	69
70	0.112	0.175	0.238	0.333	0.490	0.648	0.805	0.963	1.28	70
	3	5	7	10	15	20	25	30	40	

Figure 37. Example of a standard current-meter rating table No. 2 for Price AA current meters with cat's-whisker and magnetic contact chambers.

Electromagnetic Current Meters

Electromagnetic current meters, with no moving parts, are commercially available for measuring point velocities. These meters are based on the principle that a conductor (in this case, water) moving through a magnetic field will produce an electric current. The velocity of the moving water can be related to the electric current produced, and the distortion created in the magnetic field. The electromagnetic meters can be accurately calibrated in a tow tank, similar to the calibration of mechanical meters; however, tests have shown that the electromagnetic meters are less accurate than the Price AA meters, especially at low velocities (less than about 0.5 ft/s). The Price AA meters also have less variance than the electromagnetic meters at all velocities. Advantages of the electromagnetic current meter are as follows: no moving parts; direct readout of velocity; and, in oblique flow, the velocity measured is normal to the measuring section when the meter is held normal to it.

Marsh-McBirney 2000

An electromagnetic current meter successfully used by the USGS for making discharge measurements is the Model 2000, produced by Marsh-McBirney. This meter, as shown in figure 38, is designed to mount on a standard round or top-setting wading rod. The meter is not designed for cable suspension.

A display meter, also shown in figure 38, shows a direct readout of the velocity. No conversion equation or table is necessary. The meter must be kept clean for accurate readings, and it is recommended that the rating be occasionally spot checked to verify that it is still accurate. This can be done in two ways. First, submerge the meter in a bucket of still water to verify the zero point of the rating. Second, place the meter in close proximity to a Price AA meter, in flowing water, to verify that it gives the same velocity reading. If there are differences, rate the electromagnetic meter again in the tow tank.

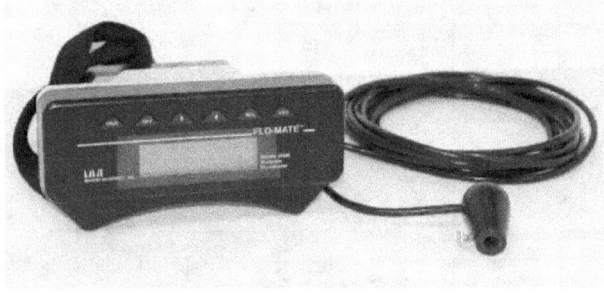

Figure 38.   Marsh-McBirney Model 2000 electromagnetic flowmeter and display meter.

Ott Electromagnetic Current Meter

An Ott electromagnetic current meter is available; however, it has not been used extensively in the United States. The Ott meter, shown in figure 39, works in a manner similar to the Marsh-McBirney meter.

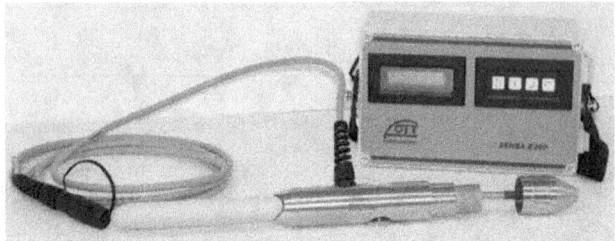

Figure 39.   Ott electromagnetic current meter.

Acoustic Current Meters

Acoustic Doppler current meters, with no moving parts, are commercially available for measuring point velocities. These meters are based on the Doppler principle. The velocity of the moving water is measured using the transmitted and received signals from sound pulses reflecting off particles in the moving water column. These acoustic meters can be accurately calibrated in a tow tank, similar to the calibration of mechanical meters. Advantages of an acoustic Doppler current meter are as follows: no moving parts; direct readout of velocity; and ability to sense very low velocities less than the rated velocities in standard mechanical current meters.

Acoustic Doppler Velocimeter (ADV)

The SonTek/YSI FlowTracker handheld ADV ("FlowTracker" and "ADV" are used interchangeably in this chapter) is designed as an alternative to the Price AA and pygmy meters for wading discharge measurements. The FlowTracker operates at an acoustic frequency of 10 MHz and measures the phase change caused by the Doppler shift in acoustic frequency that occurs when a transmitted acoustic signal reflects off particles in the flow. The magnitude of the phase change is proportional to the flow velocity. The phase difference can be positive or negative, allowing ADVs to measure positive and negative velocities. The FlowTracker measures the velocity at a rate of approximately 10 MHz, averages the data, and records 1-second velocity-vector data.

The maximum velocity the FlowTracker can measure is reduced when measuring flow that is not perpendicular to the transmitting transducer. The receiving transducers can measure a velocity range of only ±1.15 m/s (3.77 ft/s). A velocity component placed directly toward or away from the receiving transducers larger than 1.15 m/s (3.77 ft/s) will result in erroneous velocities. Because of the geometric arrangement of the transmitting and receiving transducers, a velocity of 4.5 m/s flowing perpendicular to the transmitting transducer face will result in the maximum velocity towards a receiving transducer of 1.15 m/s (3.77 ft/s).

The FlowTracker probe is mounted to a standard top-setting wading rod with a special offset-mounting bracket (fig. 40). This bracket is designed to locate the FlowTracker probe at the front of the wading rod, with the sampling volume about 2 in. (5 cm) to the right of the wading rod. Although the probe is inserted into the flow, the sampling volume is about 4 in. (10 cm) away from all physical parts of the probe, to minimize flow disturbance in the sampling volume.

FlowTrackers have several unique data-processing requirements because of their method of operation and some of the inherent limitations of the acoustic Doppler measurement technique. Unlike mechanical meters that use the momentum of the water to turn a propeller and directly measure the velocity of the water, the FlowTracker does not measure the velocity of the water. The FlowTracker measures the velocity of particles (sediment, small organisms, and bubbles) suspended in the flow, assuming that these particles travel at the same velocity as the water. Therefore, the quality of the measurement is dependent on the presence of particles within the sampling volume that reflect a transmitted signal. The FlowTracker records the signal-to-noise ratio (SNR), standard error of velocity (based on 1-second data), angle of the measured flow (relative to the x-axis of the FlowTracker probe), number of filtered velocity spikes, and a boundary quality-control flag. These velocity and quality-assurance data may be used to evaluate the measurement conditions. Few similar quality-assurance data are available for Price current-meter measurements.

Although a FlowTracker can measure within about 1.2 in. (3 cm) of a boundary, the velocity measurement might be affected by acoustic interference when the sampling volume is close to boundaries or underwater objects, even when the sampling volume is not directly on or past the boundary. At the start of each velocity measurement, if the probe detects nearby acoustic boundaries that could cause interference with the velocity measurement, a boundary adjustment is automatically made. The boundary adjustment attempts to overcome the possible interference by reducing the lag times of the acoustic signals transmitted by the FlowTracker, causing a reduction of the velocity range that can be measured. Any changes are noted in the boundary quality-control flag. Because the sampling volume is located about 4 in. (10 cm) from the transmitting transducer it can be difficult to ascertain the precise location of the sampling volume. If the sampling volume is on or past a boundary, the velocity data will be erroneous. Be careful to avoid boundaries while making measurements in depths less than 3.54 in. (9 cm), especially in channels with irregular bottoms.

Spikes in velocity data occur with any acoustic Doppler velocity sensor such as the FlowTracker. Spikes may have a variety of causes (for example, large particles in the flow, air bubbles, or acoustic anomalies). Velocity data from each FlowTracker measurement are evaluated to look for spikes. The FlowTracker spike filter is a variation on a method called "Tukey's Outlier." In this method, a histogram of each velocity component is calculated. The FlowTracker determines the lower quartile ($Q1$; 25 percent of samples are less than this value), the upper quartile ($Q3$; 75 percent of samples are less than this value), and the interquartile range ($IQR = Q3-Q1$). If the IQR is less than 0.015 m/s (0.049 ft/s), IQR is set to

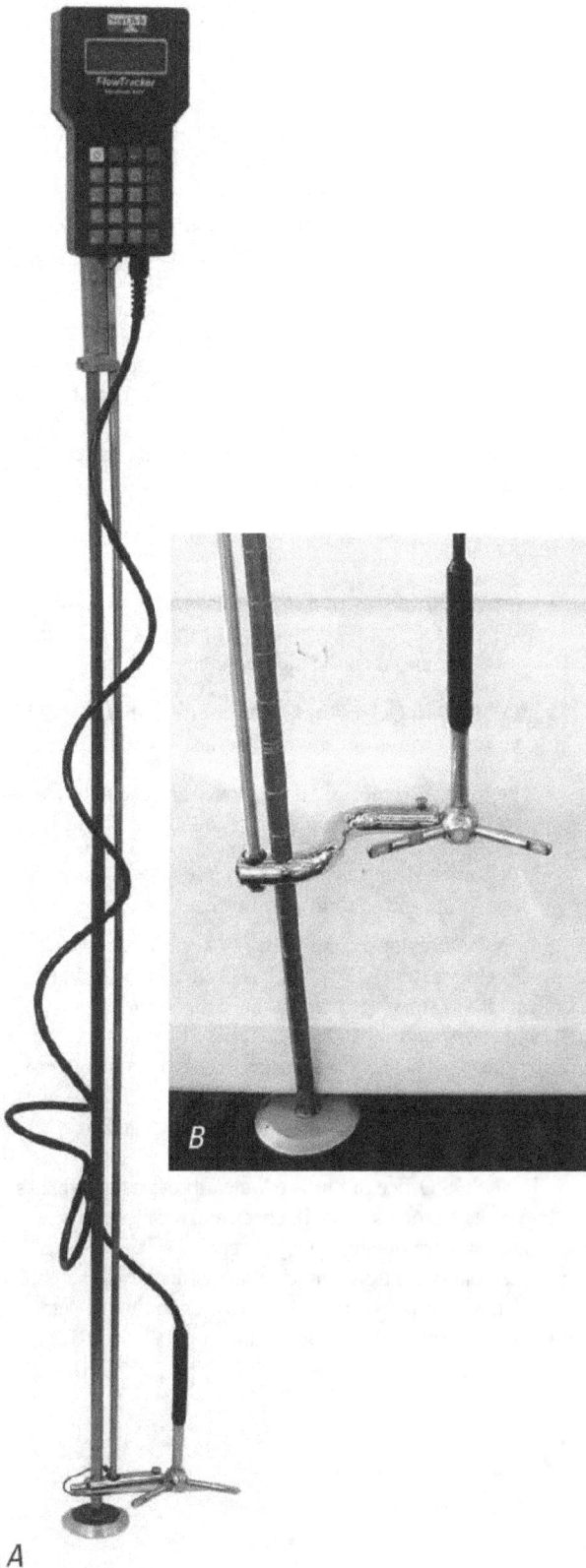

Figure 40. *A*, SonTek/YSI FlowTracker acoustic Doppler velocimeter (ADV) mounted on a standard top-setting wading rod and *B*, closer view of transmitting and receiving transducers and offset-mounting bracket.

0.015 m/s (0.049 ft/s). Any value less than ($Q1–2*IQR$) or greater than ($Q3+2*IQR$) is considered a spike and is not used for mean-velocity calculations.

The FlowTracker measures magnitude and direction of velocity. The operator must keep the wading rod perpendicular to the tag line so that the pulse generated by the transmitter is parallel to the tag line, regardless of flow direction. To compute discharge, the FlowTracker uses the component of velocity perpendicular to the transmitting transducer and reports the flow angle from the FlowTracker's x-axis as a quality-control value. A flow angle measured by the FlowTracker may be the result of flow that is not perpendicular to the tag line, or a wading rod that is not being held perpendicular to the tag line (operator error). Flow angles of less than 20 degrees with small variations between verticals are not unusual. Large fluctuations of flow angles between verticals, however, may indicate a poorly measured cross section. If there is angular flow, and the wading rod is oriented with the flow, the velocity used and resulting discharge would be biased high. If the flow is truly perpendicular to the cross section, but the wading rod is erroneously held at an angle, the velocity and resulting discharge would be biased low. To avoid possible errors in the measured velocities, it is important that the operator always carefully and accurately aligns the wading rod.

Signal-to-Noise Ratio (SNR)

Adequate signal-to-noise ratio is needed to obtain an accurate measurement of the flow velocity. SNR is a measure of the strength of the reflected acoustic signal relative to the ambient noise level of the instrument. SNR is a function of the concentration and size distribution of the particles that reflect the acoustic signal. SNR is recorded for each beam with each 1-second sample. The manufacturer states that optimal SNR is 10 decibels (dB) or above (SonTek/YSI, 2002). USGS policy is that FlowTrackers should not be used for measuring discharge if the SNR for any single beam is less than 4 dB.

Speed of Sound

The accuracy of hydroacoustics instruments like the FlowTracker is dependent on an accurate speed of sound. The speed of sound is primarily a function of the temperature and salinity of the water. The FlowTracker has a built-in temperature sensor. To verify that the temperature sensor is working correctly, take an independent water-temperature measurement prior to each discharge measurement. If the FlowTracker

has been stored in an environment with a different ambient temperature from the water, the probe may need to be placed in the water for a period of time, allowing it to equilibrate with the water temperature. A 5°F error in temperature will result in approximately a 1-percent bias in the measured velocity. The speed of sound is also sensitive to salinity. A 5-part-per-thousand error in salinity would result in an approximate velocity bias of 1 percent, when used in saline environments like estuaries; therefore, the operator needs to measure the salinity and input the value into the FlowTracker.

Maintenance and Care

Although the built-in QCTest is reliable for detecting issues, a BeamCheck stores more system performance data and still may be needed to evaluate the unit in more detail when there is a potential issue.

QCTests and BeamChecks

- Perform a QCTest and store it with each measurement. When a QCTest is completed as part of a measurement, it will print out on the measurement summary.

- Complete a QCTest in flowing water with the sample volume away from any boundaries.

- Perform a BeamCheck if you notice any anomalies in the QCTest. Any failures in a QCTest require a BeamCheck.

- Perform a BeamCheck after any possible physical damage (drop, and so forth), firmware upgrade, or repair.

As stated previously, the FlowTracker is an acoustic Doppler velocimeter (ADV) that has been adapted to fit on a typical USGS streamgaging wading rod, developed by the USGS in cooperation with the SonTek/YSI Inc., and is widely used by the USGS. The FlowTracker has undergone extensive testing to evaluate differences between the FlowTracker performance and vertical-axis current meters (that is, Price AA, pygmy, and so forth).

The USGS Office of Surface Water, through the HIF, has put into place a process that will check and recalibrate each FlowTracker approximately every 3 years to ensure the quality assurance/quality control of this instrument in the measurement of the Nation's surface-water resources. For additional details, see Office of Surface Water Memorandum 2010.02 (2010).

Acoustic Digital Current Meter (ADC)

Another development that is a potential alternative to the Price AA and pygmy meters for wading discharge measurements is the Ott acoustic digital current meter (ADC) (fig. 41). The Ott ADC operates with two transducers at an acoustic frequency of 6 MHz, and measures the phase change caused by the Doppler shift in acoustic frequency that occurs when a transmitted acoustic signal reflects off particles in the flow. The phase measurement is restricted to ±180 degrees and uses a pulse scheme with two different time delays to resolve the phase ambiguity.

The velocity (V) is computed using the following formula:

$$V = \frac{c \times \Delta\Phi}{4 \times \Pi \times \tau}, \qquad (19)$$

where
V velocity (in distance per unit time),
c speed of sound in water (in distance per unit time),
$\Delta\Phi$ computed phase difference, and
τ time lag between pulses.

The Ott ADC uses a pulse-coherent technique. Transmitted pulses have a known lag time (τ). Backscatter echoes are amplified in the sensor head and then sent to the handheld display where they are digitized. A stable quartz oscillator controls the measurement sequence.

The Ott ADC has several unique data-processing requirements because of its method of operation and some of the inherent limitations of the acoustic Doppler measurement technique. Unlike mechanical meters that use the momentum of the water to turn a propeller and directly measure the velocity of the water, the Ott ADC, as with other acoustic Doppler current meters in use in the USGS and elsewhere, does not measure the velocity of the water. The Ott ADC measures the velocity of particles (sediment, small organisms, and bubbles) suspended in the flow, assuming that these particles travel at the same velocity as the

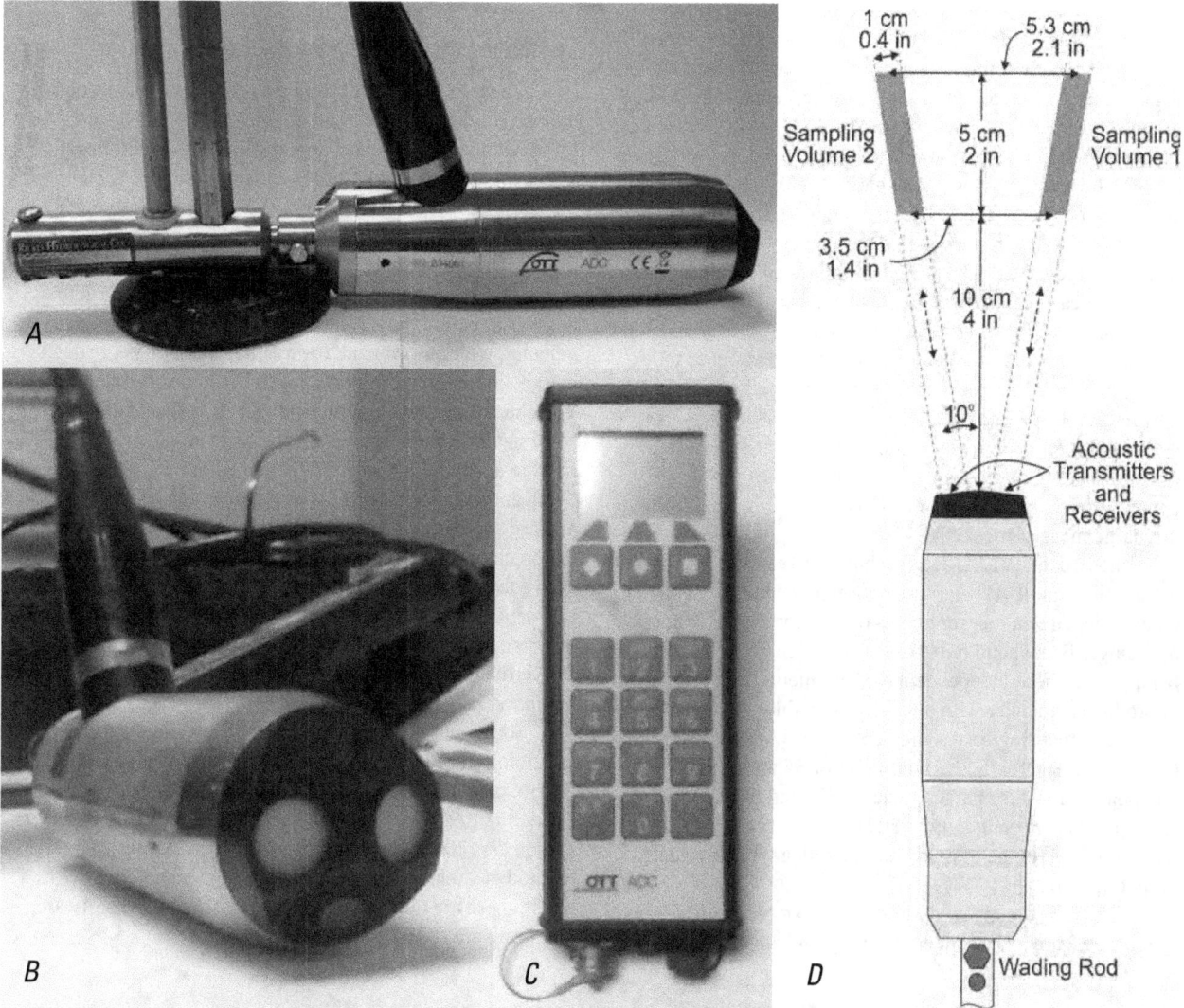

Figure 41. *A*, Ott acoustic digital current meter (ADC) mounted on a standard top-setting wading rod; *B*, closer view of transmitting and receiving transducers; *C*, ADC datalogger; and *D*, schematic of the sampling volume and transducers.

water. Therefore, the quality of the measurement is dependent on the presence of particles within the sampling volume that reflect a transmitted signal. In addition, the Ott ADC measures depth by using an absolute piezoresistive pressure cell with a range of 0 to about 16.4 ft (0–5 m). The pressure cell is located inside the probe and protected by the wading rod adapter, which is securely fastened with two screws. The hydrographer can estimate a depth without reading the top-setting rod.

Preliminary testing of the Ott ADC at the HIF indicates this technology has a lot of potential for use as an alternative to mechanical current meters.

Acoustic Doppler Current Profiler (ADCP)

The ADCP measures velocity magnitude and direction using the Doppler shift of acoustic energy reflected by material suspended in the water column, providing essentially a complete vertical profile of velocity.

Broadband, narrowband, and other pulse-to-pulse coherent ADCPs transmit pairs of acoustic pulses along a narrow beam from each of multiple transducers. As the pulses travel through the water column, they strike suspended sediment and organic particles (referred to as "scatterers") that reflect some of the acoustic energy back to the ADCP. The ADCP receives and records the reflected pulses. The reflected pulses are separated by time differences (range gating) into successive volumes called "depth cells." The frequency shift (known as the Doppler effect) is proportional to the velocity of the scatterers relative to the ADCP. The ADCP computes a velocity component along each beam, because the beams are positioned at a known angle from the vertical (usually 20 or 30 degrees) and in known horizontal orientations so that trigonometric relations can be used to compute three-dimensional water-velocity vectors for each depth cell. Therefore, the ADCP produces vertical profiles of velocity composed of water speeds and directions at regularly spaced intervals, vertical profiles of velocity, discharge profiles, and a wealth of information for a discharge measurement (fig. 42).

The Broadband ADCPs (such as the Teledyne RD Instruments Rio Grande ADCPs) use phase-coded pulses, such that many independent measurements of velocity can be made by a single Broadband pulse of the same length as a narrowband pulse. These independent measurements are averaged to produce a velocity with a lower uncertainty than would be possible with a single measurement. Narrow-band systems (such as the SonTek/YSI RiverSurveyor S5 and M9) typically compensate for this characteristic by pinging faster (sending more pulses per second, up to 20 Hz) and reporting a velocity based on the average of many pulses with a typical velocity output of 1 Hz.

Currently, the most common main external components of an ADCP are a transducer assembly and a pressure case.

Commonly, the transducer assembly consists of three to nine transducers that operate at a fixed, ultrasonic frequency, typically 300 to 3,000 kilohertz (kHz). The transducers are horizontally spaced around the transducer assembly; all transducers have the same fixed angle from the vertical, referred to as a "beam angle," that is typically between 20 and 30 degrees. The transducer assembly may have a convex or concave configuration or, in the case of the phased array, an essentially flat surface. The pressure case is attached to the transducer assembly. Examples of several different ADCPs used in the USGS are shown in figure 43.

When an ADCP is deployed from a moving boat, it is connected by cable to a power source and by cable or radio modem to a portable microcomputer. The computer is used to program the instrument, monitor its operation, and collect and store the data. For a detailed description of how an ADCP measures velocity and computes discharge, and detailed instruction in the use of ADCP technology with reference to moving boats, see Mueller and Wagner (2009).

Midsection Method With an ADCP

When an ADCP is deployed to perform a midsection method discharge measurement, most best practices used for mechanical-meter discharge measurements still apply. These practices, including site selection, are well documented in Rantz and others (1982). The midsection method with an ADCP is similar to the midsection method used for mechanical-meter discharge measurements and involves measuring the channel area and water velocities of a stream at a cross section; however, instead of point measurements of velocity, with this method, the average velocity is obtained by profiling the velocity in the water column at each section. The channel is divided into a number of vertical subsections. Most natural channels must be divided into 20 or 30 subsections to adequately characterize their irregular geometry. The depth and average velocity are measured at each subsection and are applied to a subarea whose width extends halfway to the preceding and following observation points. The area of each subsection is determined by directly measuring width and depth. The average water velocity in each subsection is estimated using the measured velocity at elected locations in the vertical. The total discharge within the stream is the sum of the individual subsection discharges.

Additional details and in-depth discussion of ADCP technology, methodology, and quality assurance and quality control can be found in USGS publications by Simpson and Oltmann (1993), in Lipscomb (1995), Morlock (1996), Oberg and others (2005), and Mueller and Wagner (2009). The ADCP method is a relatively new and evolving technology, and as a result, there are ongoing changes to the hardware, software, and firmware.

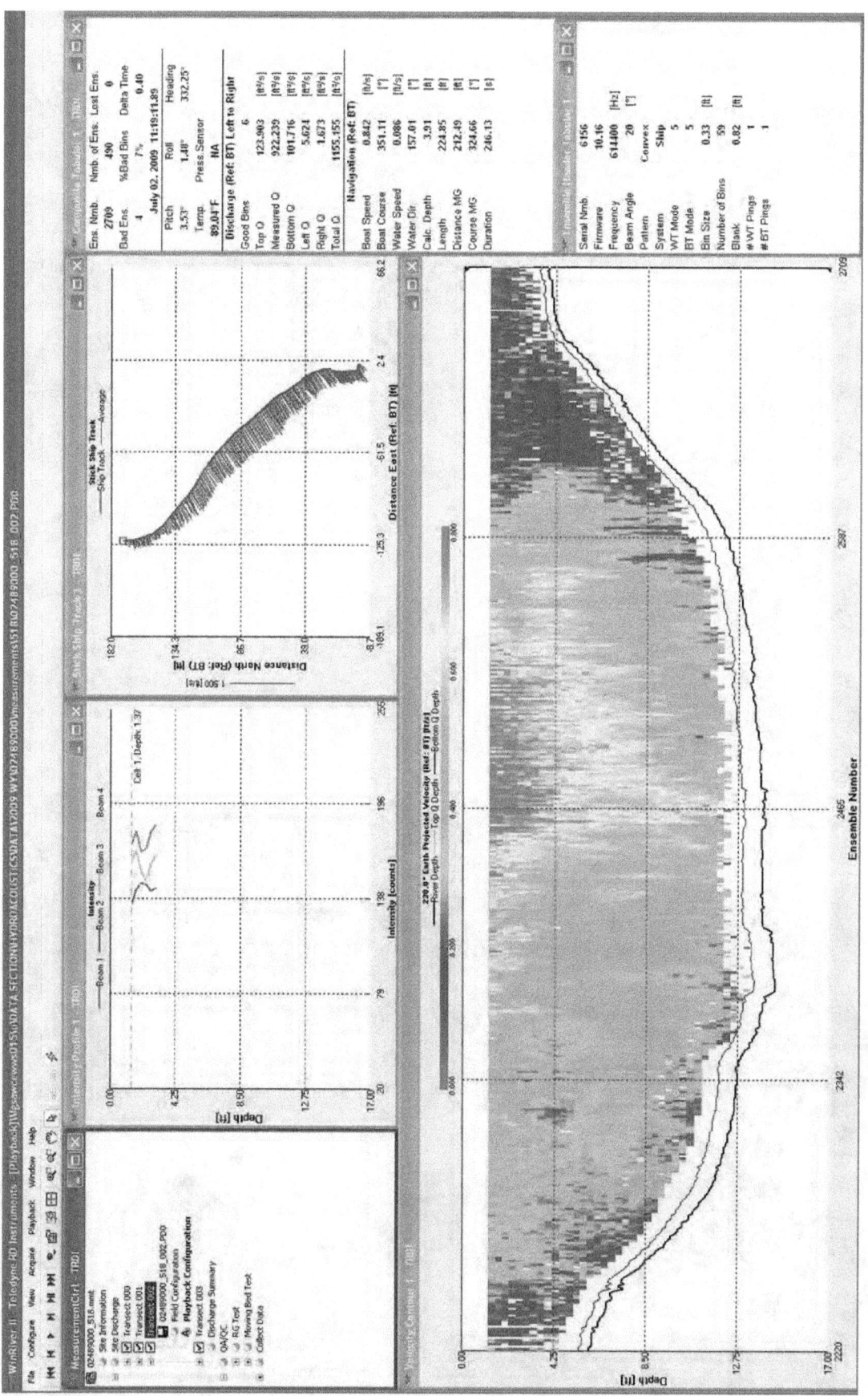

Figure 42. Screen capture of an ADCP measurement of discharge for the Pearl River near Columbia, MS (02489000).

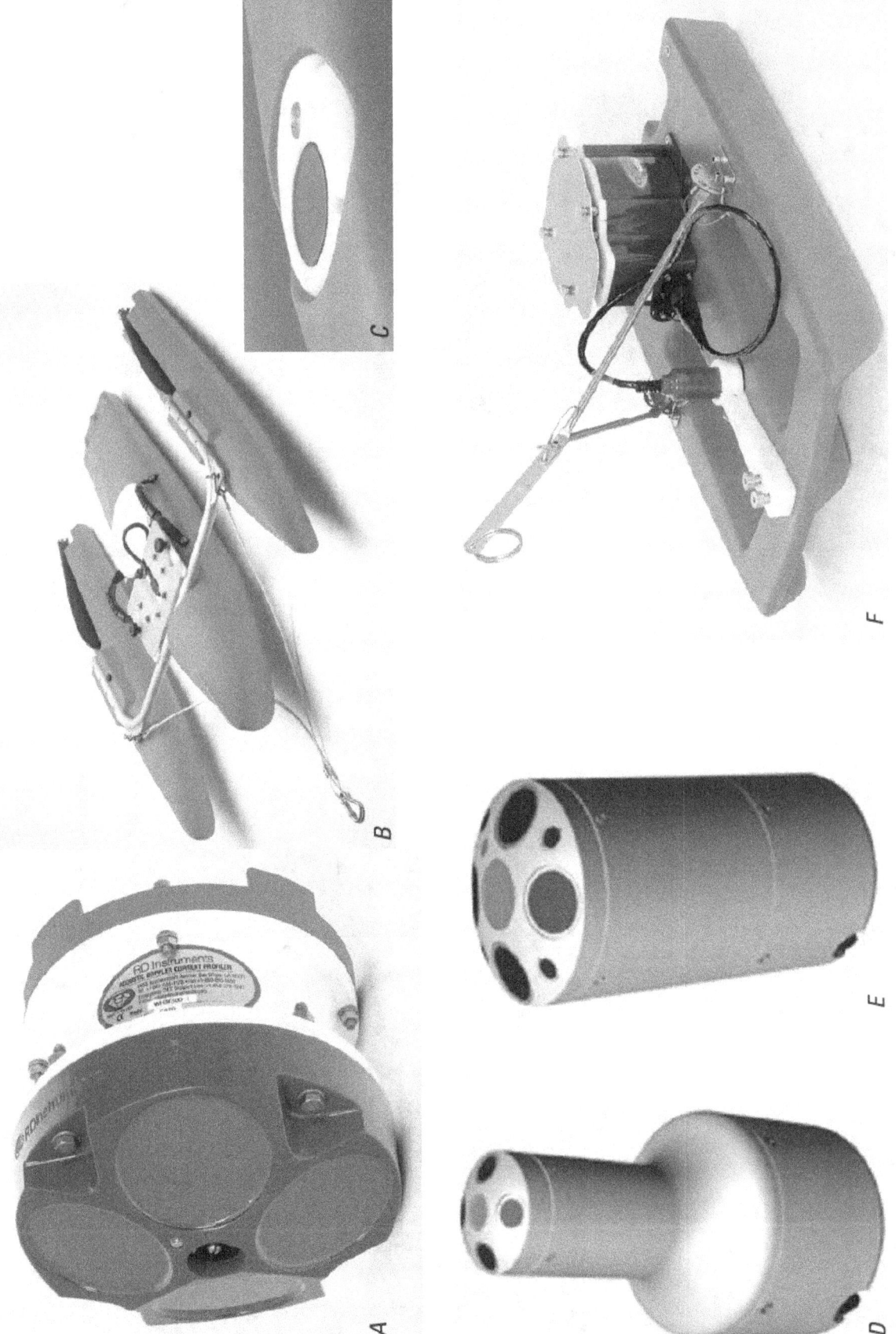

Figure 43. Examp es of acoust c Dopp er current profi ers (ADCPs) used to measure d scharge at gag ng stat ons; *A*, Te edyne RD Instruments 600 kHz R o Grande ADCP; *B*, Te edyne RD Instruments 600 kHz R verRay phased array ADCP, and *C*, c oseup v ew of phased array transducer; *D*, SonTek/YSI R verSurveyor S5; *E*, SonTek/YSI R verSurveyor M9; and *F*, Te edyne RD Instruments StreamPro ADCP

Optical Current Meters

An optical current meter, as shown in figure 44, is a stroboscopic device designed to measure surface velocities in open channels without immersing equipment in the stream. The optical current meter is used principally in measurements of surface velocity during floods when it is impossible to use streamgaging equipment that requires placement in the water, because of extremely high velocities and high-debris content in the stream. Use of this technology and other technologies, such as radar, deserves further investigation in the measurement of discharge using surface velocities.

Figure 44.   Hydrographer using an optical current meter to measure surface velocity.

Sounding Equipment

Sounding (determination of depth) is commonly done mechanically; the equipment used depends upon the type of measurement being made. Measure depth and position in the vertical with a rigid rod or by a sounding weight suspended from a cable. The cable is controlled either by a reel or by a handline. A sonic sounder is also available, but it is usually used in conjunction with a reel and a sounding weight. The various equipment that can be used for sounding is described in the following paragraphs. In addition, ADCPs can sound depths using sophisticated algorithms that may have global positioning features and capabilities. Mueller and Wagner (2009) discuss ADCP sounding in more detail.

Top-Setting Wading Rods

The two types of wading rods commonly used are the top-setting rod and the round rod. The top-setting rod is preferred because of the convenience in setting the meter at the proper depth and because the hydrographer can keep his hands dry.

The top-setting wading rod, as shown in figure 45, has a ½-in. hexagonal main rod for measuring depth and a ⅜-in. diameter round rod for setting the position of the current meter.

The rod is placed in the stream so the base plate rests on the streambed, and the depth of water is read on the graduated main rod. When the setting rod is adjusted to read the depth of water, the meter is positioned automatically for the 0.6-depth method, as shown in figure 46. The 0.6-depth setting is the setting measured down from the water surface. This setting is the same as the 0.4-depth position when measured up from the streambed. When the depth of water is divided by 2, and this value is set on the setting rod, the meter will be at the 0.8-depth position from the water surface. When the depth of water is multiplied by 2, and this value is set, the meter will be at the 0.2-depth position from the water surface.

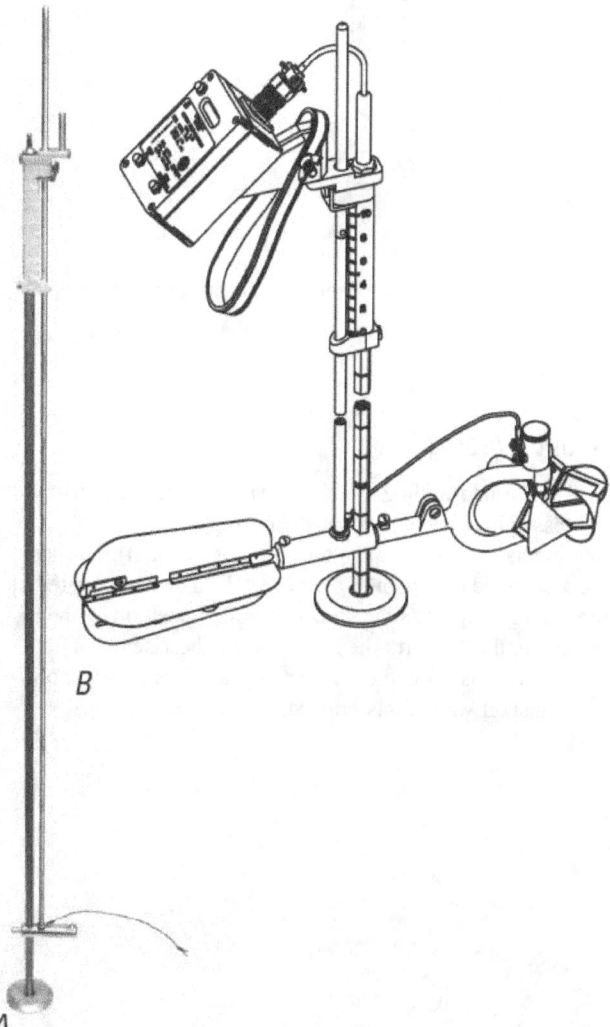

B

A

Figure 45.   *A*, Top-setting wading rod and *B*, schematic of a top-setting wading rod with Price AA current meter and current-meter digitizer (CMD) attached.

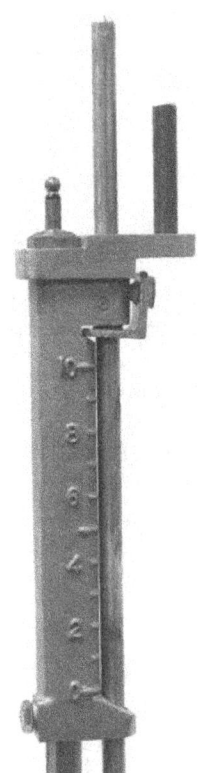

Figure 46. Closeup view of setting scale on handle of top-setting wading rod.

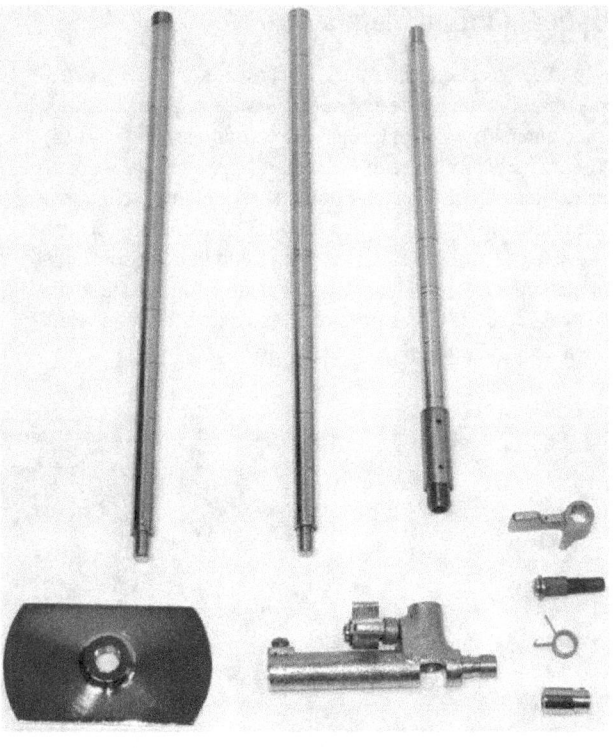

Figure 47. Parts for the round wading rod.

Round Wading Rods

The round wading rod, as shown in figure 47, consists of a base plate, lower section, sliding support, three or four intermediate sections, and a rod end (not essential). The parts are assembled as shown in figure 48. The meter is mounted on the sliding support and is set at the desired position on the rod by sliding the support. The round rod can be assembled into various lengths using the 1-ft sections, and it is easy to store and transport when disassembled.

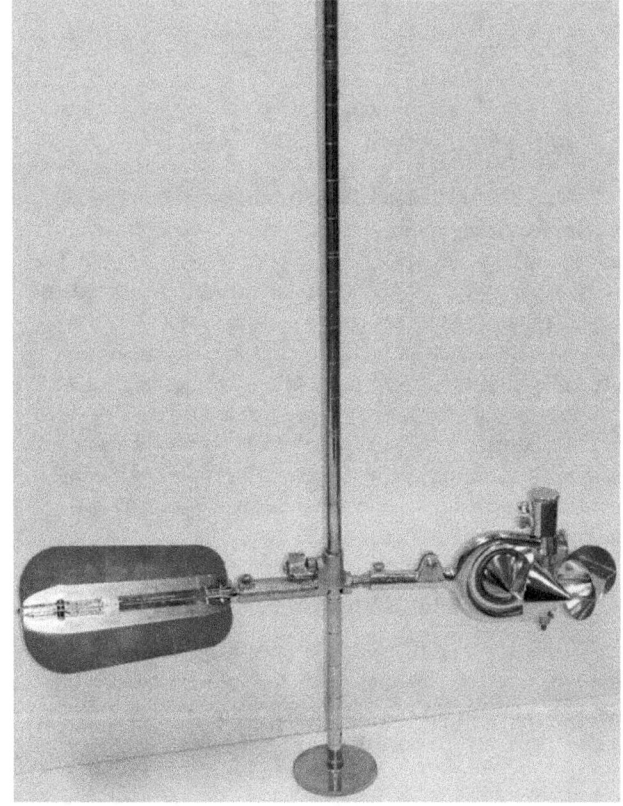

Figure 48. Round wading rod with Price AA current meter attached.

Winter-Style Suspension Wading Rods

Measurements made under ice cover should use the WSCan winter sounding rods, either in the ½- or 1-in.-diameter versions. These rods are available in sections so that the desired length can be assembled. A special foot fits the lower section, and the rods will accommodate the winter-style current-meter yoke, as shown in figure 49.

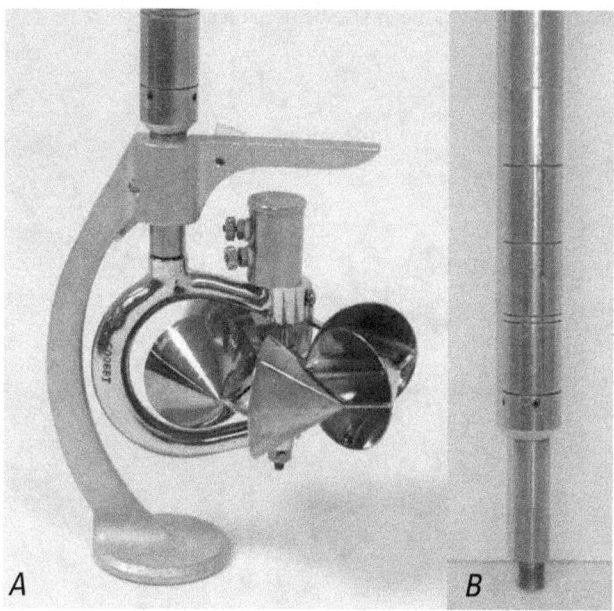

Figure 49. Water Survey Canada winter-style round 1-inch suspension rod and meter: *A*, with Price AA meter in a winter-style yoke attached and *B*, closer view of round 1-inch diameter suspension rod.

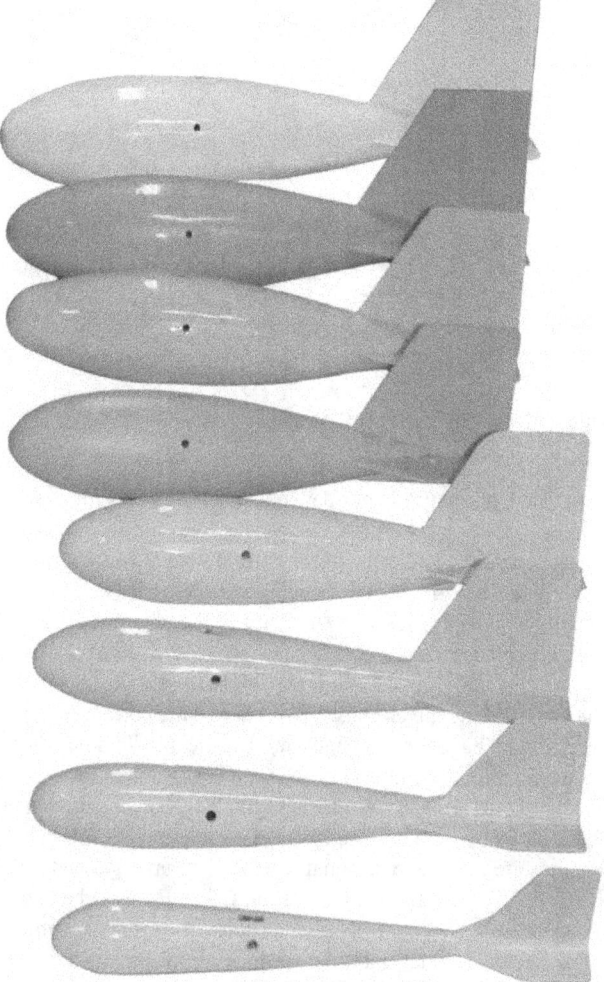

Figure 50. Columbus C-type sounding weights (15 through 300 pounds).

Sounding Weights

If a stream is too deep or too swift to wade, the current meter is suspended in the water from a boat, bridge, or cableway. A sounding weight is suspended below the current meter to keep it stationary in the water. The weight also prevents damage to the meter when the assembly is lowered to the streambed.

The sounding weights currently used are the Columbus weights, commonly called the C type, and are shown in figure 50. The weights are streamlined to offer minimum resistance to flowing water. The weights are available in 15-, 30-, 50-, 75-, 100-, 150-, 200-, and 300-pound sizes. Each weight has a vertical slot and a drilled horizontal hole to accommodate a weight hanger and securing pin.

Hanger Bars

The weight hanger is attached to the end of the sounding line by a connector. The current meter is attached to the hanger bar beneath the connector, and the sounding weight is attached to the lower end of the hanger bar.

There are three types of weight hanger bars, as shown in figure 51:

1. The Columbus or C type, ⅛ × ¾ × 12 in. (for weights up to 150 pounds);

2. Heavy weight, ⅛ × ¾ × 18 in. (for 200- and 300-pound weights);

3. Heavy weight, ⅛ × 1 ½ ×18 in. (for 200- and 300-pound sounding weights that have the slots properly extended to accommodate a 1½-in. wide hanger bar).

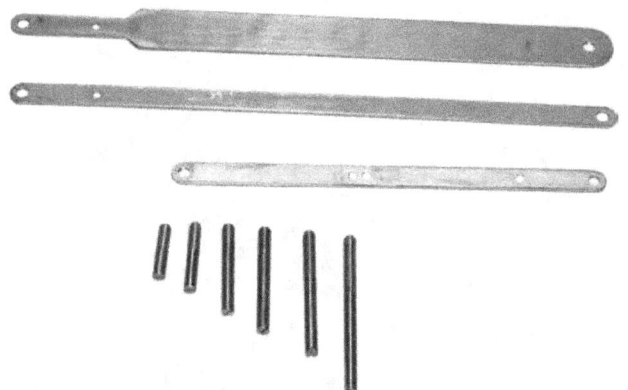

Figure 51.   Sounding weight hanger bars and hanger pins.

The Columbus hanger bar contains three holes in order to properly position the meter. The hanger screw of the current-meter yoke is placed through the bottom hole to support the meter when a 30-pound sounding weight is used. The center of the meter cups is then 0.5 ft above the bottom of the weight. This arrangement is designated as 30 C .5, which means that a 30-pound Columbus weight is being used and the center of the meter cups is 0.5 ft above the bottom of the weight. The hanger screw goes through the middle hole when 15- or 50-pound weights are used. The designations for these arrangements are 15 C .5 and 50 C .55. The hanger screw goes through the upper hole when 50-, 75-, 100-, and 150-pound weights are used. The designations for these arrangements are 50 C .9, 75 C 1.0, 100 C 1.0, and 150 C 1.0. Each of the two heavy-weight hangers has only one hole for the hanger screw of the meter. The designations for these arrangements are 200 C 1.5 and 300 C 1.5.

Weight-hanger pins of various lengths, as shown in figure 51, are available for attaching the sounding weight to the hanger bar. The stainless steel pins are threaded on one end to screw into the hanger bar and slotted on the other.

Sounding Reels

Several different types of sounding reels are available for use with the Columbus C-type weights. In general, a sounding reel has a drum for winding the sounding cable, a crank-and-ratchet assembly for raising and lowering the weight or holding it in any desired position, and a depth indicator. Table 11 contains detailed information on each of the five most commonly used reels.

The A-pack reel, as shown in figure 52, is light, compact, and ideal for use at cableway sites a considerable distance from the highway. It can also be used on cranes, bridge boards, and boat booms.

Figure 52.   A-pack reel.

Table 11.   Sounding reel data.

Reel	Sounding cable	Cable diameter, in inches	Drum circumference, in feet	Cable capacity, in feet	Maximum weight, in pounds	Depth indicator	Brake	Type of operation
A-pack	Ellsworth	0.084	1	45	50	Counter	No	Hand.
Canfield	Single conductor[1]	.0625	1	45	50	Counter	No	Hand.
A-55	Ellsworth	.084 .10	1	95 80	50 100	Self computing	No	Hand.
B-56	Ellsworth	.10 .125	1.5	144 115	150 200	Self computing	Yes	Hand or power.
E-53	Ellsworth	.10 .125	2	206 165	150 300	Self computing	Yes	Power.

[1]Some Canfield reels have been converted to double-conductor cable but most of them are still used as single-conductor reels.

The Canfield reel, as shown in figure 53, is also compact with uses similar to that of the A-pack reel. The Canfield reel is not available from the HIF, and must be obtained from Leupold and Stevens Instruments, Inc.

The A–55 reel is for general purpose use with the lighter sounding weights, as shown in figure 54.

The B–56 reel (a major modification of the B–50 reel) can handle all but the heaviest sounding weights and has the advantage that it can be used with a hand crank or power equipment, as shown in figure 54.

The E–53 reel is the largest reel commonly used for current-meter measurements. This reel will handle the heaviest sounding weights and is designed exclusively for use with power equipment. It has a hand crank for emergency use, as shown in figure 55.

Figure 53. Canfield reel.

Figure 55. E–53 reel.

Sounding Cable

Ellsworth reverse-lay two-conductor stranded cable is normally used on all sounding reels, except the single-conductor Canfield reel, which uses galvanized steel aircraft cord. Ellsworth cables are normally available in 0.084-, 0.100-, and 0.125-in. diameters. It is important to use the appropriate size cable-laying sheave on the reels.

For safety purposes, when measuring floods, it is important to connect the sounding cable to the sounding reel in such a way that the cable will break in the event that heavy debris is caught and cannot be released. The cable will usually unwind (pay out) from the sounding reel until it reaches its end, at which point there is danger to the equipment and the hydrographer unless the cable is cut or breaks. Precut some of the cable strands when installing the cable on the reel so that the remaining strands will break when the load reaches a specified limit. Table 8 provides information about cable strength and number of strands to cut to provide the necessary safety margin. Complete instructions for making the cable installation are given in a HIF technical information sheet dated April 1999.

A

B

Figure 54. *A*, A–55 reel and *B*, B–56 reel.

Connectors

A connector is used to join the end of the sounding cable to the sounding-weight hanger. The three types of connectors generally used are types B, C1, and pressed sleeve, as shown in figure 56. The type B connector is used with A–55, B–56, and E–53 reels. The C1 connector is used with the A-pack and Canfield reels, although the pressed-sleeve connector can be used on these reels. The pressed-sleeve connector is also used on handlines.

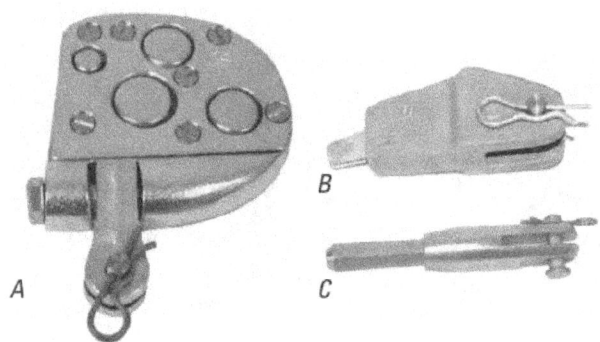

Figure 56. Connectors for attaching sounding cable to sounding-weight hanger; *A*, Type B; *B*, Type C1; and *C*, pressed sleeve.

Depth Indicators

A computing depth indicator, as shown in figure 57, is used on the A–55, B–56, and E–53 reels. The stainless-steel indicator is less than 3 in. in diameter and has nylon bushings that do not require oil. The main dial is graduated in feet and tenths of a foot from 0 to 10 ft. The depth is indicated by a pointer. Tens of feet are read on a numbered inner dial through an aperture near the top of the main dial.

The main dial has a graduated spiral to indicate directly the 0.8-depth position for depths up to 30 ft.

The A-pack and Canfield reels, shown in figures 52 and 53, are equipped with counters for indicating depths.

Figure 57. Computing depth indicator.

Handlines

A handline, shown in figure 58, is a device used for making discharge measurements from bridges using a 15- or 30-pound sounding weight. The advantages of using the handline are that it is easy to set up, it eliminates the use of a sounding reel and supporting equipment, and it reduces the difficulty in making measurements from bridges, which have interfering members, such as trusses. The disadvantages of using the handline are that there is a greater possibility of making errors in determining depth because of slippage of the handline, measuring scale, or tape, and it requires great physical exertion, especially in deep streams. Handlines can be used from cable cars, but this is not recommended because of the disadvantages mentioned above.

Ellsworth cable is recommended for handlines because of its flexibility and durability. Two-conductor electrical service cord is used between the headset connector and the handline reel.

The pressed-sleeve connector or the C1 connector is used on handlines because they are lighter in weight than the type B connector, yet strong enough for the sounding weights used with handlines.

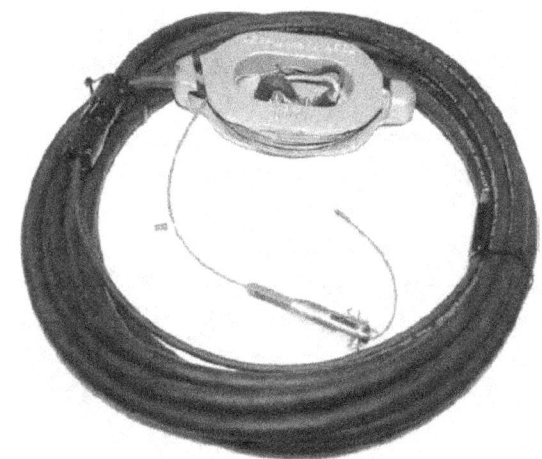

Figure 58. Handline.

Power Unit

Power units, as shown in figure 59, are available for the B–56 and E–53 reels to raise and lower the sounding weight and meter. The power unit can be used with 6-, 12-, 18-, or 24-volt batteries.

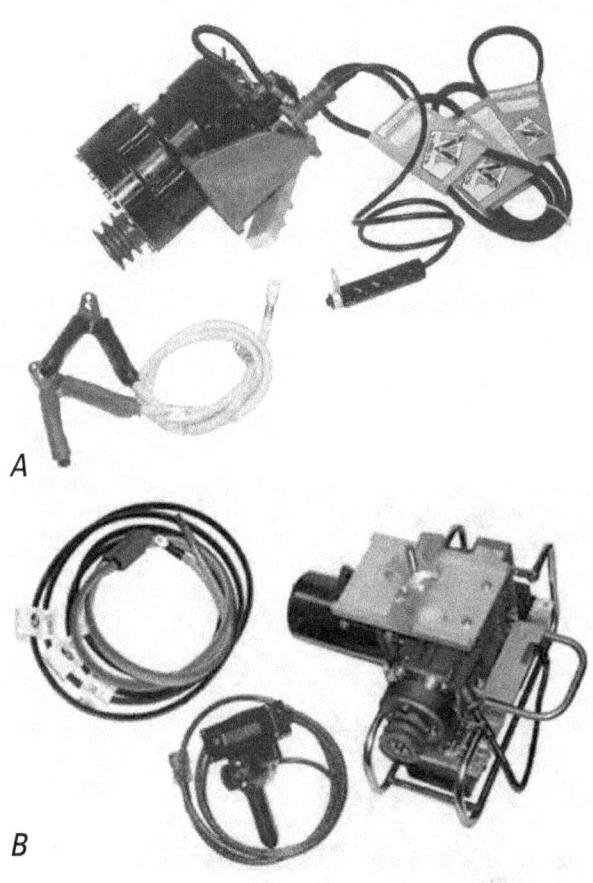

A

B

Figure 59. Power units for sounding reels; *A*, J & H Single Speed Power Drive and *B*, USGS Variable Speed Reel Drive System.

Sonic Sounder

A commercial, compact, portable sonic sounder has been adapted to measure stream depth. The sounder is powered by either a 6- or 12-volt storage battery and will operate continuously for 10 hours on a single battery charge. Three recording speeds are available—36, 90, and 180 in. per hour. Four operating ranges—0 to 60 ft, 60 to 120 ft, 120 to 180 ft, and 180 to 240 ft—allow intervals of 60 ft of depth. The sounder is portable, weighing only 46 pounds. The depth recorded is from the water surface to the streambed. The transducer has a narrow beam angle of 6 degrees, which minimizes errors on inclined streambeds and allows the hydrographer to work close to piers or other obstructions.

Measurements can be made with this equipment without lowering the meter and weight to the streambed. As soon as the weight is in the water, the depth will be recorded. The meter can then be set at the 0.2 depth, or just below the water surface where a velocity reading is obtained. Then a coefficient is applied to convert measured velocity to the mean in the vertical.

Temperature change affects the sound propagation velocity, but this error is limited to about plus or minus 2 percent in freshwater. This error can be eliminated completely by adjusting the sounder to read correctly at a particular average depth determined by other means.

Cableway Equipment

The USGS cableway provides a track for the operation of a manned cable car from which the hydrographer makes a current-meter measurement. Most cableways have a clear span of 1,000 ft or less, although a few structures have been built with clear spans approaching 2,000 ft. The design and construction of cableways are described in detail by Wagner (1995).

Cable cars provide a movable platform from which the hydrographer, sounding reel, and other necessary equipment are supported. The newer versions of these cable cars are fabricated from aluminum, and have a standard follower brake and integral-reel mounts, which will accept all standard sounding reels, including battery-powered reels. Cable cars can also be equipped with the Sandpoint type cable-car brake, which allows the cable car to be slowed or stopped. Both sitdown and standup types of manually propelled cable cars are used in streamgaging, as shown in figures 60 and 61, and have space for two people to work. Some older cable cars still in use are fabricated partially from wood, may or may not have permanent reel mounts, and may have space for only one person.

Manned cable cars are moved from one point to another on the cableway by means of cable-car pullers, as shown in figure 62. The standard car puller is an aluminum-cast handle with a snub attached. The snub, usually four-ply belting, is placed between one of the car sheaves and the cable to prevent movement of the car along the cable. A second type of puller, also shown in figure 62, is used when a car is equipped with a follower brake. A third type, the Colorado River cable-car

Figure 60.   Sit-down cable car.

Figure 61.   Stand-up cable car.

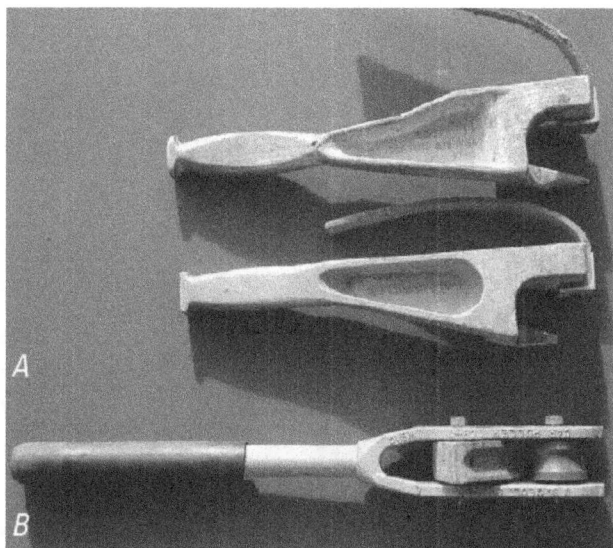

Figure 62. Cable-car pullers: *A*, with belt and *B*, with follower brake.

puller, is the same in principle as the puller used on cars equipped with a follower brake.

Power-operated cable cars, such as the battery-powered car shown in figure 63, are available for extremely long spans or other special situations where extensive streamgaging and monitoring is required. The power assist on these cable cars is also utilized to operate a type E sounding reel.

Unmanned, remotely operated cable carriages, such as the Hydrological Services Hornet, are used for discharge measurements as well as for sediment sampling. The cable carriage and sounding equipment can be remotely operated from the stream bank, as shown in figure 64. They are used in areas where it is impossible to wade, where no bridges are available, and where it is not practical to build or maintain a manned cableway.

Unmanned cableways are used more in Europe than in the United States, but are becoming more common in the United States. These bank-operated cableways have obvious benefits in safety and convenience (fig. 65). Both permanent and portable bank-operated cableways are becoming more useful in the measurement of discharge, especially for more narrow streams, generally 300 ft or less in top-bank width.

Figure 63. Battery-powered cable car.

Figure 64. Remotely operated Hydrological Services Hornet cable carriage with ADCP and trimaran used by the USGS.

Figure 65. Bank-operated cableway.

Bridge Equipment

Streamflow measurements are frequently made from a bridge. The meter and sounding weight can be supported by a handline, a bridge board, or by a sounding reel mounted on a crane. An ADCP mounted on a tethered craft can also be deployed from a bridge. Tethered ADCP craft are rapidly becoming the prevalent means of measuring discharge from a bridge.

Handlines and Bridge Boards

A handline, as described in a previous section of this chapter, is the simplest form of bridge-measuring equipment. Used extensively in the 20th century, it does not require any separate reels or handling equipment; however, it can only be used with light sounding weights, such as the 15- and 30-pound size. It also requires that depth be measured with tags, and a tape or measuring stick.

A bridge board is a portable platform made from wood or metal upon which a small reel can be mounted. Bridge boards may be used with an A-pack, A–55, or B–56 sounding reel and weights up to 75 pounds. A bridge board is usually about 6 to 8 ft long, with a sheave at one end over which the meter cable passes, and a reel seat near the other end. The board is placed on the bridge rail so that the force exerted by the sounding weight suspended from the reel cable is counterbalanced by the weight of the sounding reel. The bridge board may be hinged near the middle to allow one end to be placed on the sidewalk or roadway. Figure 66 shows a bridge board in use.

Figure 66.   Measuring from a bridge with a bridge board.

Portable Cranes

Two types of hand-operated portable cranes are the type A for weights up to 100 pounds, and the type E for weights greater than 100 pounds. The type A crane mounts on a three-wheel or four-wheel base or truck, and the type E crane mounts on a four-wheel base or truck. Cranes can be easily moved by hand along the sidewalk or floor of the bridge. Figure 67 shows a type A crane mounted on a three-wheel base, and figure 68 shows a type E crane mounted on a four-wheel base.

Any of the reels described in table 11 may be used on either of the portable cranes; however, the power-driven reels (B–56 and E–53) are used only with the Type E crane. Various combinations of cranes, bases (trucks), and reels are possible.

Figure 67.   Type A crane mounted on a three-wheel base.

Figure 68.   Type E crane mounted on a four-wheel base.

All cranes are designed so that the crane can be tilted forward over the bridge rail far enough for the meter and weight to clear most rails and be lowered to the water. Where bridge members obstruct passage of the crane along the bridge, the weight and meter can be raised and the crane can be tilted back to pass by the obstruction.

Use cast-iron counterweights weighing 60 pounds with four-wheel-base cranes. The number of such weights needed depends upon the size of sounding weight being supported, the depth and velocity of the stream, and the amount of debris being carried by the stream.

Use a protractor on the outer end of cranes to measure the angle the sounding line makes with the vertical when the weight and meter are dragged downstream by high-velocity water. The protractor is a graduated circle clamped to an aluminum plate. A plastic tube, partly filled with colored antifreeze (ethylene glycol), is the protractor index. This tube is fitted in a groove between the graduated circle and the aluminum plate. A stainless-steel rod is attached to the lower end of the plate to ride against the downstream side of the sounding cable. The protractor will measure vertical angles from -25 degrees to +90 degrees. Figure 69 is a close-up view of a protractor mounted at the outer end of the boom.

Figure 69. Protractor used for measuring vertical angles.

Power-Driven Cranes

Many special arrangements for measuring from bridges have been devised to suit a particular purpose. Vehicle-mounted cranes are often used for measuring from bridges over larger rivers, as shown in figure 70. Monorail streamgaging cars have also been developed for large rivers. The car is suspended from the substructure of bridges by means of I-beams.

Tethered Craft

The USGS, in cooperation with manufacturers, continues to test and refine tethered-platform designs for measuring streamflow. Platform specifications have been developed for

Figure 70. Vehicle-mounted, power-driven crane.

radio-modem telemetry of acoustic Doppler current profiler (ADCP) data, potential platform-hull sources have been investigated, and many hull configurations have been tested and evaluated.

Platforms, which included a variety of hull configurations, were tested for drag and stability at the USGS Hydraulic Laboratory tow tank and at a field site below a reservoir. The testing indicated that trimaran designs provided the best all-around performance under a range of conditions. The trimaran designs house the ADCP in the center hull. Waterproof radio modems that operate at 900 MHz are used to communicate wirelessly with instruments at high-baud rates.

A tethered-platform design with a trimaran hull, and 900-MHz radio modems, are commercially available from several vendors. Continued field use has resulted in USGS procedures for making tethered-platform discharge measurements, including methods for tethered-boat deployment, moving-bed tests, and measurement of edge distances. Figure 71 shows a tethered craft ADCP in wide usage in the USGS (Mueller and Wagner, 2009; Rehmel and others, 2003).

Figure 71. Measuring with a tethered ADCP, DGPS, and trimaran.

Boat Equipment

There are five basic types of boat measurements: the manual stationary boat, the manual moving boat, the automatic moving boat, the ADCP moving boat, and the remote-controlled ADCP moving boat. Equipment requirements for each of these boat measurement types are described in the following paragraphs.

Manual Stationary Boat

The manual stationary boat method uses a boat as a platform for the hydrographer and the sounding equipment. The boat is attached to a tag line or cable to stabilize it at each vertical where soundings are made. The heavy-duty tag lines required for boat measurements are described in a previous section of this chapter.

Special equipment assemblies, as shown in figure 72, are necessary to suspend the meter from the boat if the depths do not allow using rod suspension. A crosspiece spanning the boat is clamped to its sides, and a boom attached to the center of the crosspiece extends out over the bow. The crosspiece is equipped with a guide sheave and clamp arrangement at each end to attach the boat to the tag line, and makes it possible to slide the boat along the tag line from one station to the next. A small rope can be attached to these clamps so that in an emergency a tug on the rope will release the boat from the tag line. The crosspiece also has a clamp that prevents lateral movement of the boat along the tag line during readings. The boom consists of two structural aluminum channels, one telescoped within the other to permit adjustments in length. The boom is equipped with a reel plate on one end and a sheave over which the meter cable passes. The sheave end of the boom is designed so that by adding a cable clip to the sounding cable a short distance above the connector, the sheave end of the boom can be retracted when the meter is raised out of the water. The raised meter is easy to clean and is in a convenient position when not being operated.

Figure 72. Manual stationary-boat equipment assembly.

All sounding reels fit the boat boom, except the A-pack and the Canfield, which can be made to fit by drilling additional holes in the reel plate on the boom.

In addition to the equipment already mentioned, the following items are needed for making boat measurements:

1. A stable boat big enough to support the hydrographers and equipment;

2. A motor that can easily move the boat against the maximum current in the stream;

3. A pair of oars for standby use;

4. A personal floatation device (PFD) for each hydrographer; and

5. A bailing device.

Manual Moving Boat

Equipment requirements for the manual moving boat method of making a discharge measurement are described in detail by Smoot and Novak (1969); they are not described in this chapter. In summary, the manual moving-boat method requires a sonic sounder, a vane with indicator, a special current meter, and an easily maneuverable small boat. The manual moving-boat method is seldom used since the advent of the ADCP moving-boat method, which is described below.

Automatic Moving Boat

The automatic moving-boat method is similar to the manual moving-boat method, except that all readings of depth, velocity, angles, and distance are recorded automatically by onboard computer equipment. Equipment requirements are more complex in order to enable the automatic sensing and recording.

Like the manual moving-boat method, the automatic moving-boat method is seldom used since the advent of the ADCP moving-boat method.

ADCP Moving Boat

Currently, the ADCP moving-boat method is the most common moving-boat method in the USGS. All readings of depth, velocity, angles, and distance are recorded to a laptop or PDA and the discharge measurement is computed using ADCP software and input from the hydrographer.

As shown in figure 73, described in other sections of this chapter, and in depth by Mueller and Wagoner (2009), ADCP moving-boat measurement techniques have almost entirely superseded the automatic moving method.

Remote-Controlled ADCP Moving Boat

There are commercially available unmanned, remote-controlled craft with ADCPs used by the USGS and others for the measurement of discharge where a manned boat or tethered boat may not be feasible. Similar to but smaller than the ADCP moving-boat, these remote-controlled craft typically come with self-contained motors and a remote-controlled system for driving the boat across a stream or river. See Mueller and Wagner (2009) for an in depth discussion of remote-controlled ADCP moving boat discharge measurements.

Figure 73. ADCP equipment mounted and operated on *A*, a moving boat and *B*, a tethered platform.

Electronic Field Notebooks and Personal Digital Assistants (PDAs)

Recent developments in electronics have produced commercially available electronic field notebooks (EFNs) and personal digital assistants (PDAs), designed specifically for the purpose of recording field notes while making a discharge measurement (fig. 74) . The PDAs are also designed for other field procedures, including discharge-measurement data collection and processing, station inspection, differential-level survey notes, and water-quality data collection and equipment calibration. The USGS commonly uses the Surface Water Measurements and Inspections (SWAMI) program with a PDA. The program can be used to record discharge measurements, inspections, differential-level surveys, and other field measurements. This software provides an efficient means of collecting field data (fig. 2C) and has been specifically designed to interface with the USGS NWIS.

The recording process is semiautomatic. Basic information and data still must be entered manually for measurements of stream depth, stationing, horizontal angle of flow, equipment, and other basic data.

EFNs automatically count meter revolutions and elapsed times, and make the conversion to stream velocity. They also assist the hydrographer with certain tasks, such as locating each subsection so that no more than 10 percent of the total flow will be included in each subsection. The notebook makes all of the measurement calculations to obtain the final discharge, and to summarize all pertinent items of the measurement.

The EFNs and PDAs can store many discharge-measurement reports. The EFN and (or) PDA reports for a discharge measurement are similar to the paper note sheets used for manual note keeping. A header, similar to the paper "front sheet," contains site information, equipment information, and a summary of measurement data. The report also contains

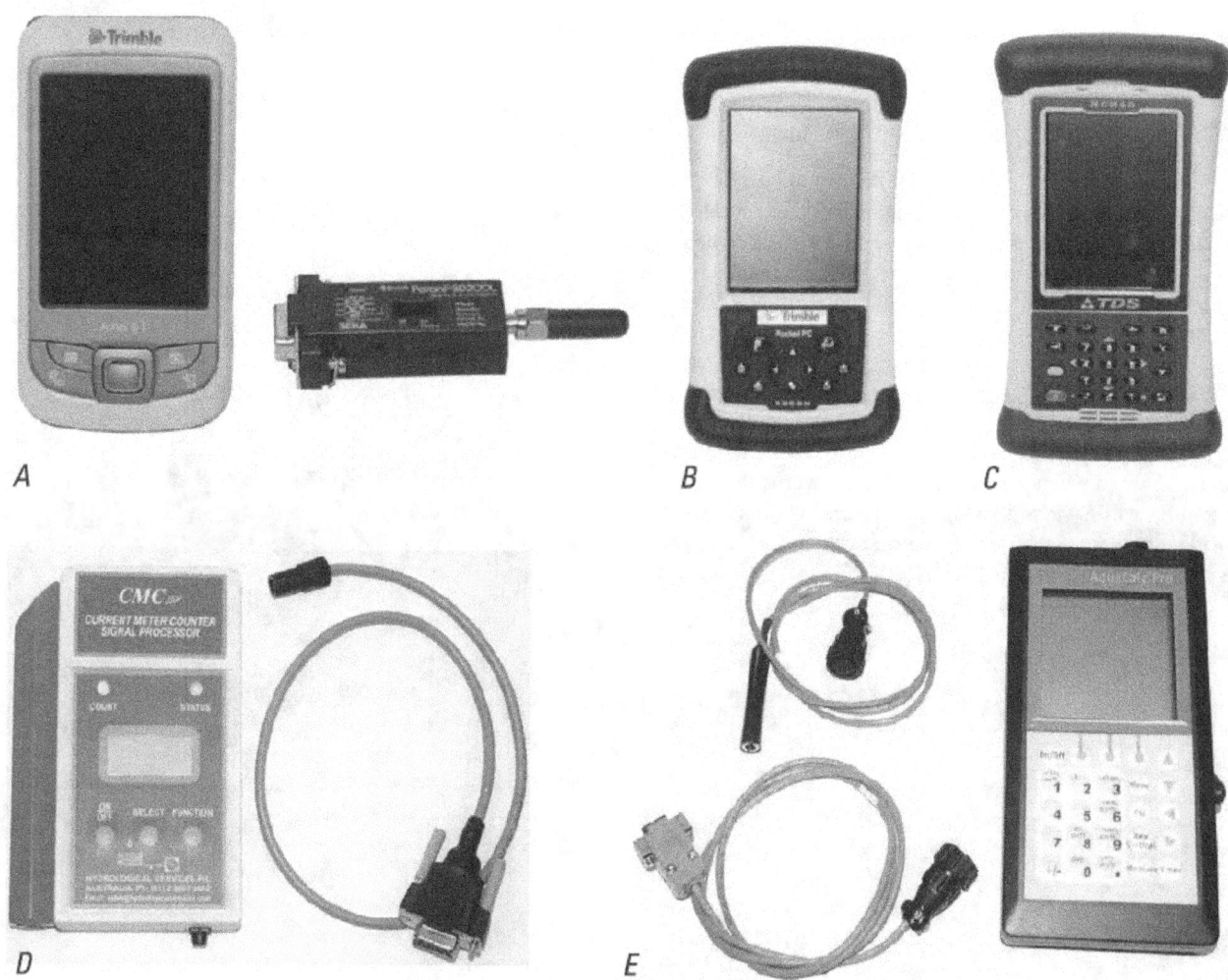

Figure 74.   Personal digital assistants (PDAs) and electronic field notebooks (EFNs), clockwise from the top: *A,* the Trimble Juno SB PDA with GPS, Bluetooth, and Parani Serial Adapter; *B,* the Trimble Nomad 800 GL PDA with GPS and Bluetooth; *C,* the Trimble Recon PDA without GPS; *D,* the JBS Instruments Aquacalc Pro Discharge Measurement Computer; and *E,* the Hydrological Services Current Meter Counter Signal Processor with Bluetooth (CMCsp).

complete measurement data, similar to the paper "inside notes," for all of the individual subsections. In addition, the report contains various warning flags and quality-control information. Complete reports for each discharge measurement can be downloaded to a computer for viewing, printing, and other analysis.

One EFN widely used within the USGS is the Aquacalc Pro Discharge Measurement Computer (Aquacalc) by JBS Instruments. The Current Meter Counter Signal Processor (CMCsp) by Hydrological Services is used in concert with a PDA to measure discharge with a mechanical current meter, as shown in figure 74. The Aquacalc and the CMCsp will work with the cat's-whisker-, magnetic-, and optic-contact chambers on the Price current meters. As with all EFNs, care should be taken to avoid false counting of meter revolutions when measuring low velocities with mechanical meters that have a cat's-whisker-contact chamber.

PDAs are manufactured by many computer companies in the United States and abroad. Several are considered reliable and have been successfully field tested. At this writing, the PDA can interface with most of the ADCPs and electronic data loggers used by the USGS.

Miscellaneous Equipment and Personal Items

Other personal equipment and items will be needed while making discharge measurements, or when working in and around rivers, creeks, and streams. Waders or boots should be worn while making wading measurements. Waders should be loose fitting, even when allowing for heavy winter clothing.

Ice chains, as shown in figure 75, should be strapped onto the soles of boots or waders for use on steep or icy stream banks, and on rocky or smooth and slippery streambeds.

A properly fitted personal-floatation device (PFD) must be worn when working near, in, or over water. This includes while wading streams, and working on a cableway, bridge, or water retention or control structure, on ice, or in a boat.

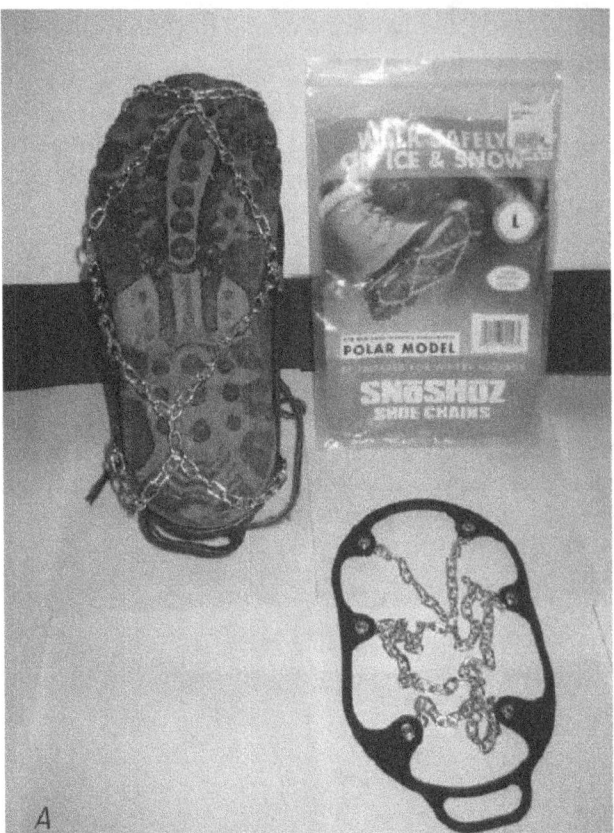

Figure 75. *A*, Ice chains for boots and waders and *B*, Sure Grip Ice Treads for boots and waders.

Accuracy of Current-Meter Discharge Measurements

The accuracy of a discharge measurement is dependent on many factors, including the equipment used, the location and characteristics of the measuring section, the number and spacing of verticals, the rate of change in stage, the measurement of depth and velocity, presence of ice and (or) debris in the measuring section, wind, experience of the hydrographer, carefulness (or carelessness) of the hydrographer, and various conditions that can occur during the process of making the measurement. The evaluation of the accuracy of a measurement has long been a qualitative assessment that takes some or all of these factors into account. A quantitative measure of the accuracy for some discharge measurements can also be made. The following two sections of this chapter describe these methods.

Qualitative Evaluation

Every discharge measurement should be evaluated for accuracy using the qualitative method. Historically, this has been the preferred method, and the hydrographer should make this evaluation immediately after making the measurement. The evaluation should be based on the hydrographer's opinion of the accuracy of the measurement—not on how well, or how poorly, the measurement plots on the stage-discharge relation. It is difficult to provide written guidelines for making a qualitative evaluation of accuracy. A good qualitative evaluation depends mostly on the experience and training of the hydrographer. Several of the factors that should be considered by the hydrographer are as follows:

Measuring section.—Consider factors such as the uniformity of depths, the smoothness of the streambed, the streambed material (that is, smooth sand; small, firm gravel; large rocks; soft muck; and so forth), the ability to accurately measure the depth, the approach conditions, presence of bridge piers, and other conditions that would affect measurement accuracy.

Velocity conditions.—Consider smoothness of velocity, uniformity of velocity, very slow velocity, very high velocity, turbulence, obstructions that may affect the vertical velocity distribution, use of one-point or two-point method, length of counting (40 or more seconds versus half-counts), and other factors that affect accuracy of velocity measurements.

Equipment.—Consider the type of current meter used (Price AA, Price pygmy, acoustic, or electromagnetic), the type of depth-sounding equipment, and the condition of the equipment.

Spacing of observation verticals.—Use about 25 to 30 verticals for a discharge measurement, spaced so that no more than 5 percent of the total discharge is contained in each subsection. Although this is frequently difficult to attain, except in unusual cases, no more than 10 percent of the total discharge should be in a subsection. Otherwise, the accuracy will be negatively affected.

Rapidly changing stage.—Although discussed in previous sections of this chapter, this condition should also be considered when assessing the accuracy of the measurement. Using the shortcut methods previously described will result in less accurate measurements of discharge.

Ice measurements.—Making discharge measurements under ice cover is usually more difficult, and sometimes less accurate, than making open-water discharge measurements. Presence of slush ice, layered ice, and anchor ice will have adverse affects on accurate measurement of depth and velocity. Velocity distribution will be affected if the water surface is in contact with the ice. Freezing of water in the meter cups and pivot chamber may affect performance of the equipment.

Wind.—Wind can affect the accuracy of a discharge measurement by obscuring the angle of the current, or by creating waves that make it difficult to sense the water surface prior to making depth soundings. Wind can also affect the vertical-velocity distribution, particularly near the surface, and can cause vertical and (or) horizontal movement of the current meter while making a boat measurement, introducing possible error in velocity measurements.

The qualitative method of assessing the accuracy of a discharge measurement requires that the hydrographer consider all of the above items and their cumulative effect on the measurement accuracy. The front page of the discharge measurement note sheet (see figure 2) has space for describing (1) the cross section, (2) the flow, (3) the weather, and (4) any other flow conditions that relate to the accuracy. These descriptions, along with the type of equipment, number of verticals, velocity measurement method, and other measurement conditions, should provide the basis for rating the measurement as excellent (2 percent), good (5 percent), fair (8 percent), or poor (more than 8 percent).

For instance, a measurement might be rated as excellent (2 percent) if (1) the cross section is smooth, firm, and uniform; (2) the velocity is smooth and evenly distributed; (3) the equipment is in good condition; (4) the two-point velocity measurement method was used; and (5) weather conditions are good (no wind or ice). On the other hand, if several of these factors make it difficult to accurately measure depth and (or) velocity, the measurement might be rated fair (8 percent), or even poor (more than 8 percent).

As stated previously, it is not possible to provide absolute guidelines for making the qualitative evaluation of accuracy. As a general rule, the accuracy of most discharge measurements will be about 5 percent, or qualitatively a "good" measurement. This is sometimes used as the base-line accuracy, with accuracy upgraded to "excellent" when measuring conditions are substantially better than average, and accuracy downgraded to "fair" or "poor" when conditions are substantially worse than average. The qualitative-accuracy evaluation is based on the hydrographer's judgment. For more detailed qualitative-evaluation information on discharge measurements using ADCPs, see Oberg and others (2005) and Mueller and Wagner (2009).

Quantitative Evaluation

A quantitative-accuracy evaluation can be made for some current-meter discharge measurements by using the procedure described by Sauer and Meyer (1992), Herschy (1994), and the International Organization for Standardization (1997). These procedures compute the uncertainty, or standard error, using a root-mean-square error analysis of individual component errors. The component errors include errors in the measurement of width, depth, and velocity, and in computation procedures. These procedures can be used to compute the standard error for most discharge measurements made with the vertical-axis, cup-type current meter. These procedures do not apply to measurements made with other types of current meters, or other methods of making discharge measurements. Likewise, they do not apply to discharge measurements where wind, ice, boundary effects, flow obstructions, improper equipment, incorrect measuring procedures, and hydrographer carelessness are factors in the measurement.

The details of the Sauer and Meyer (1992) method are described in USGS Open-File Report 92–144, and therefore are not included in this chapter. A computer program is available to compute the standard error for individual discharge measurements, and it is recommended that this quantitative evaluation be made for each discharge measurement for which it applies. Computations using this method show that the standard error of individual discharge measurements can range from about 2 percent for ideal conditions, to about 20 percent for very poor measuring conditions. Standard errors range from about 3 percent to 6 percent for measurements having generally normal measuring conditions. The standard errors computed by this method are in close agreement with qualitative evaluations.

ADCP Discharge-Measurement Accuracy

There are many sources of error in an ADCP discharge measurement. A complete measurement is composed of the ADCP-measured channel subsection, extrapolated top subsection, extrapolated bottom subsection, and edge-estimated subsections.

The largest and most substantial subsection is the ADCP-measured channel subsection. Most errors can be greatly reduced if factors, such as moving bed, water temperature, salinity, cross-section choice, instrument configuration, and boat speed, are carefully considered and accounted for, as described in previous sections of this chapter. Software is usually provided by the manufacturer that can be used to compute the ADCP instrument error for the measured subsection.

Errors for the extrapolated top, bottom, and edge subsections will vary, depending upon the extrapolation methods and relative proportion of the total discharge represented in these subsections. Again, these errors can be kept to a minimum through proper choice of cross section and careful measurement of variables, such as ADCP transducer depth and distances from each shore to the nearest ADCP section.

Studies by Morlock (1996) and Oberg and Mueller (2007) concluded that ADCP discharge measurements can be used successfully for streamflow data collection under a variety of field conditions. In Morlock (1996), 31 ADCP discharge measurements were compared to discharge ratings defined by conventional methods for the period over which the ADCP measurements were made. These comparisons showed that 25 ADCP measurements were within 5 percent of the conventional measurements. Six of the ADCP measurements differed by more than 5 percent, the maximum departure being 7.6 percent.

The study by Morlock (1996) stated that ADCP discharge measurement error was indicated by the standard deviations of the ADCP discharge measurements. The standard deviations ranged from about 1 to 7 percent of the measurement discharges. The estimated error of each ADCP discharge measurement also was computed from formulas derived by the manufacturer of ADCPs. The computations of estimated measurement error assume that ADCP instrument- and unmeasured-subsection extrapolation errors are the main source of measurement error. The standard deviations for most ADCP discharge measurements were higher than the estimated measurement errors, indicating that significant components of measurement error were not related to the instruments; errors of this nature include temporal variations of flow. It was concluded that measurement precision can be positively affected by selection of a measurement location with minimal flow variations, and negatively affected by instrument- and boat-operation factors.

Uncertainties in Discharge Measurements

All discharge measurements, no matter how carefully made, are subject to uncertainty. The measurement uncertainty can be thought of as a quantitative measure of the dispersion of the measured discharge about the true discharge. This uncertainty arises because each measurement is subject to errors of unknown magnitude. The total uncertainty in a discharge measurement may arise from several sources, including:

- uncertainty in the measurement of the cross-sectional area, which in turn arises from the following:
  ◦ uncertainty in measurements of width; and
  ◦ uncertainty in measurements of depth;
- uncertainty in the measurement of the water-velocity profile, which in turn arises from the following:
  ◦ instrument uncertainty;
  ◦ pulsation and turbulence in open-channel flow;
  ◦ deviation from our assumptions about the vertical-velocity distribution; and
  ◦ uncertainty due to oblique angles in the flow velocity;
- uncertainty due to deviation from assumptions used in the computation procedures; and
- other random or systematic errors.

These component uncertainties can be combined to estimate the total uncertainty of a single discharge measurement. Where feasible, values for these component uncertainties should be estimated independently for each site.

The uncertainty is often expressed as a standard deviation. If we assume that measurement errors are normally distributed, then this uncertainty can be used to construct confidence intervals for the measured discharge value. For example, the true discharge can be expected to be within one standard deviation of the measured value at the 68-percent confidence level. At the 95-percent confidence level, the true discharge can be expected to be within two standard deviations of a single measured value.

Quality Assurance and Quality Control

It should be the goal of each hydrographer to make discharge measurements of the highest quality and with as little error as possible. As explained in other sections of this chapter there are many actions that must be performed before, during, and after the actual measuring process. In the many implicit decisions that must be made during the course of a discharge measurement, the hydrographer, through training and experience, must develop a keen sense of what is correct and incorrect through hydrologic/engineering judgment, and strive to continually take the correct course of action in making a discharge measurement. This is commonly known as quality assurance and quality control, sometimes referred to as QA/QC. Some of the QA/QC functions are implicit; that is, they are generally understood, performed automatically, and are not specifically defined in the measurement notes and sometimes must be accomplished through hydrologic/engineering judgement. Careful regard for safety, good hydrologic/engineering judgment, and observance of proper procedure are implicit functions that cannot be overstressed in making a precise and accurate discharge measurement. On the other hand, some actions are explicit, such as performing regular spin tests of current meters, or making check measurements when the first measurement may be suspect. Following are some of the QA/QC actions that should be observed for making high-quality discharge measurements. These are not all inclusive, and each hydrographer should always include and document any other actions that relate to the quality of the measurement. Additional QA/QC requirements are given in the QA/QC plan for each USGS Water Science Center.

- *Care of current meters, current profilers, and sounding equipment.*—Previous sections of this chapter describe the proper care of current meters, current profilers, and sounding equipment. Current meters are especially susceptible to damage and misalignment while in use, as well as in transit, if they are not properly protected. The hydrographer should follow all established guidelines to ensure that the streamgaging equipment, especially the current meter and (or) profiler, are in good working condition. While making a discharge measurement, the current meter should be periodically observed and checked to be sure it is operating smoothly and has not become fouled by debris, ice, or other obstructions.

- *Spin tests of current meters.*—One of the requirements for maintaining and checking current meters is a periodic, timed spin test under controlled conditions. The procedure for making a timed spin test is described in a previous section of this chapter. In addition, before, during, and after a discharge measurement, check that the rotor is turning smoothly and does not come to an abrupt stop.

- *Carefulness, good judgment, and proper procedure.*—It is the hydrographer's responsibility to apply proper procedures with care and good judgment while making streamflow measurements. These implicit functions of QA/QC should be observed at all times.

- *Computing and plotting the measurement on site.*—Compute a discharge measurement as soon as possible after it is completed. Do this at the site before leaving. If the measurement does not plot within 5 percent (or other specified percentage) of the rating curve in use, or if it is not in line with the previous trend of measurements, try to find an explanation. For instance, there may be an obvious change of the control that would explain the deviation. All such explanations should be documented in the measurement notes. If a satisfactory explanation cannot be found, then make a check measurement.

- *Making check measurements.*—If possible, while making a check measurement, select a different cross section from the original section and use a different current meter. Make the check measurement as close in time and gage height to the original measurement as possible.

- *Checking discharge measurements.*—In general, hand-computed discharge measurements are not checked for mathematical errors. Nevertheless, check measurements that do not plot within an acceptable percentage of the rating curve, or within the previous trend of measurements. Likewise, check measurements that define a significant extrapolation of the low end or high end of a rating curve. Discharge measurements recorded in an electronic notebook, such as the Aquacalc, are automatically computed and do not require checking.

- *Documentation of QA/QC.*—Document in the measurement notes, if possible, all measures taken to ensure that discharge measurements are accurate and of high quality. Some QA/QC measures require specific documentation independent of the measurement notes. For example, current-meter spin tests have specific forms that document the spin-test results and all repairs to the meter.

Safety Requirements

Practicing personal and overall safety is of utmost importance when working near, in, and above water. It is not the purpose of this chapter to describe all of the safety requirements; however, each hydrographer should be familiar with, and should observe, those requirements. Other documents provide the details of specific safety requirements for making wading measurements, and for working on ice, bridges, cableways, and boats. For instance, each USGS Water Science Center flood plan addresses such things as one-person versus multiple-person field parties, use of an approved PFD, and traffic control while making discharge measurements from bridges, cableway safety, and boat safety. USGS WRD Memorandum No. 99.32 (1999) provides safety guidance as it is related to discharge measurements, sampling, and other related streamgaging activities. Each Water Science Center has a safety officer and a safety plan; both should be consulted for specific safety issues.

Portable Weir-Plate Measurements

Current-meter measurements made in shallow depths and low velocities are usually inaccurate, if not impossible, to obtain. Under these conditions, a portable weir plate is a useful device for measuring the discharge.

A 90-degree V-notch weir is suitable because of its favorable accuracy at low flows. A weir made of 10- to 16-gage galvanized sheet iron will produce a free-flowing nappe, having the effect of a sharp-crested weir, and will give satisfactory performance. The thickness of the plate should vary with the size of the weir. Refer to figure 76 for recommended proportions. Decreasing the plate thickness on larger weirs will help maintain portability. The notch is cut, without sharpening, leaving a flat, even edge. Framing, in the form of small-angle irons, is required for medium and large sizes. Canvas attached on the downstream or upstream side prevents leakage under or around the weir. Eyebolts, properly placed, will secure rods driven in earth channels to stabilize the plate.

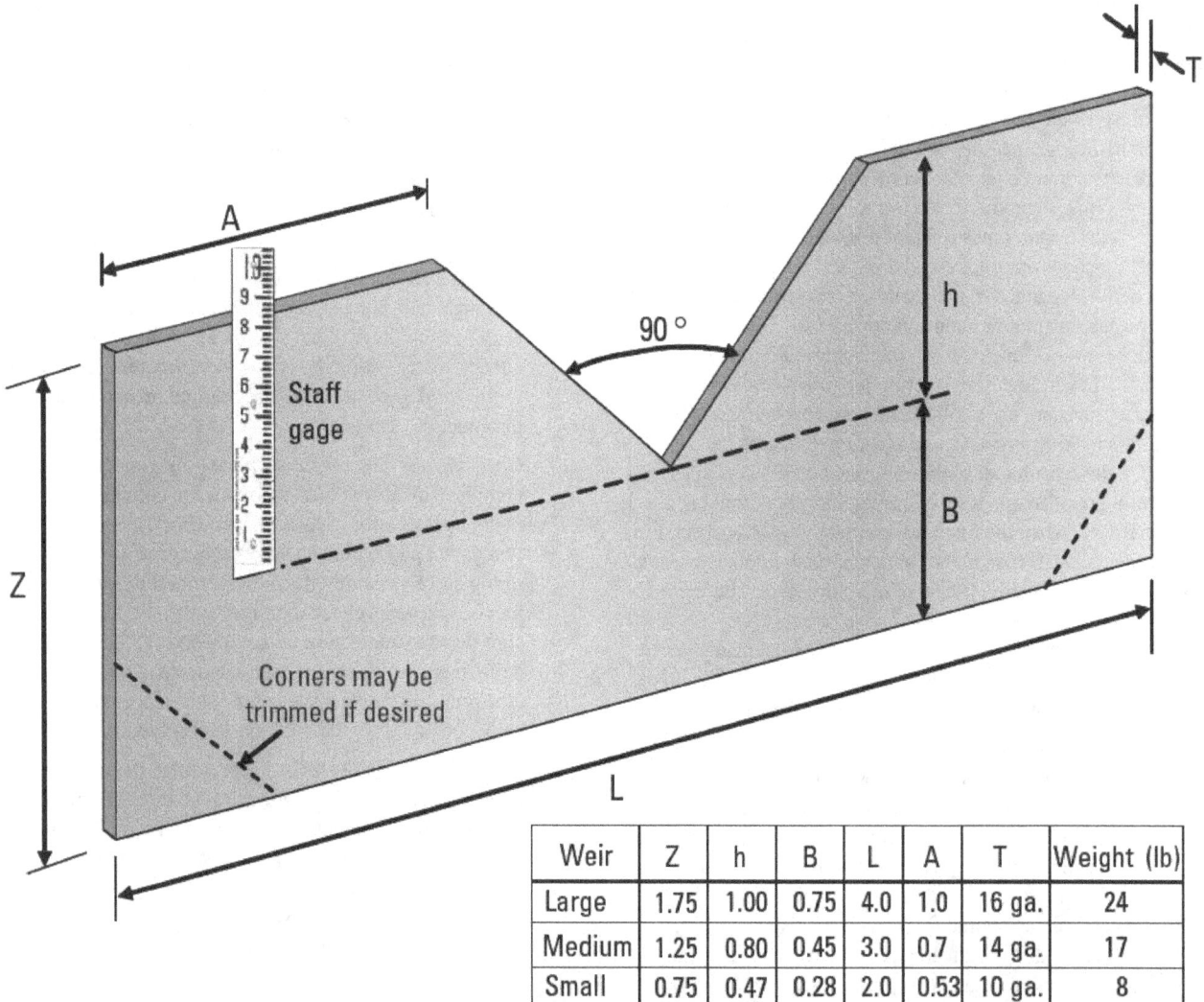

Weir	Z	h	B	L	A	T	Weight (lb)
Large	1.75	1.00	0.75	4.0	1.0	16 ga.	24
Medium	1.25	0.80	0.45	3.0	0.7	14 ga.	17
Small	0.75	0.47	0.28	2.0	0.53	10 ga.	8

Figure 76. Portable weir plate.

Attach a staff gage to the upstream face of the weir plate, with the zero point at the same elevation as the bottom of the weir notch. The staff should be far enough from the notch to be outside of the zone of drawdown, which is a distance greater than twice the head on the notch. The staff gage is used to obtain head on the weir.

The general equation for flow over a sharp-edged triangular weir with a 90-degree notch is

$$Q = Ch^{\frac{5}{2}}, \tag{20}$$

where Q discharge, in ft³/sec,
 h static head, in ft, and
 C the coefficient of discharge.

The weir should be rated by determining the flow volumetrically for various values of head, or by having it rated in the HIF laboratory. In the absence of a rating, a value of C of 2.47 may be used.

Flows from 0.02 to 2.0 ft³/s are measured with the large weir of figure 76. Discharges can be measured within 3-percent accuracy if the weir is not submerged. A weir is not submerged when there is free circulation of air on all sides of the nappe.

To place the plate in a sand or silt channel, only a carpenter's level and a shovel are needed. Push the weir into the streambed, and drive the rods through the eyebolts on each end to stabilize the weir. Use the level to make the top of the plate horizontal and the plate plumb. Another way to level the plate is by fastening a staff gage or level bubble to each end of the weir, where the staff gages are set at the same elevation. The plate is leveled by making the staff-gage readings identical or by using the level bubbles. Pack soil and streambed material around the ends and bottom of the weir to prevent leakage. Place canvas immediately downstream from the weir to prevent the falling jet from undercutting the streambed. Let the flow stabilize before making a measurement. Read the gage height at half-minute intervals for a period of about 3 minutes, and use a mean value in the above equation to compute the discharge. Ordinarily, one person can measure with a weir of this type. Remove the weir after you have completed the measurement.

Portable Parshall-Flume Measurements

A portable Parshall measuring flume is useful for measuring discharge when the depths are shallow and the velocities are low. The standard Parshall flume has a converging section, a throat, and a diverging section. The floor of the converging (or upstream) section is level both longitudinally and transversely when in place. The floor of the throat section slopes downward and the floor of the diverging or downstream section slopes upward. The standard Parshall flume can be used to measure discharge under free-flow conditions, as well as submerged conditions.

The flume used by the USGS is a modified version of the standard Parshall flume. The modification consists primarily of the removal of the downstream section, which reduces the weight of the flume and makes it easier to install. Because it has no downstream section, however, it can only be used to measure free-flow conditions (that is, where the submergence ratio is 0.6 or less). This can usually be accomplished by building up the streambed by a couple of inches under the level, converging floor of the flume when the flume is installed.

Free flow occurs when the ratio of the lower head to the upper head is less than 0.6. The discharge under this condition depends only on the length of crest (width of throat section) and depth of water at the upper gage. A flume that is properly constructed has an accuracy of 2 to 3 percent under free-flow conditions.

Install the flume by placing it in the channel; fill in with available channel bed or bank material around it to prevent any water from bypassing it. Use a carpenter's level to set the floor of the converging section level. Some flumes are equipped with levels attached to the braces on the flume. After the flume is in place, the streamflow is allowed to stabilize before reading the gage. After the flow stabilizes, take gage readings at about half-minute intervals for about 3 minutes. Use an average of the gage readings with the flume rating to determine the discharge. Remove the flume after the measurement is complete.

A modified 3-in. Parshall flume is shown in figure 77. This modified version is virtually the same as the standard Parshall flume except that it does not have a diverging section. The gage height, or upstream head on the throat, is read in the small stilling well that is hydraulically connected to the flow by a ⅜-in. hole.

The basic rating equation for a flume is

$$Q = Cbh^{\frac{3}{2}}, \tag{21}$$

where Q discharge, in cubic feet per second (ft³/s),
 C a dimensionless coefficient of discharge that can vary with head and other factors,
 b width of the throat section, in feet, and
 h head, or gage height, in the converging section, in feet.

Figure 77. Modified 3-inch Parshall flume.

The rating for the 3-in. modified Parshall flume described in this section is given in table 12, and was taken from Buchanan and Somers (1969). An identical table is published by Kilpatrick and Schneider (1983), and also by Rantz (1982). The original source of this rating is unknown, but was probably based on laboratory tests. The rating in table 12 plots as a straight line on logarithmic plotting paper, and the equation for this rating was computed by regression analysis as follows:

$$Q = 1.1392h^{1.5797}. \qquad (22)$$

This equation is close to one that can be derived for the 3-in. modified Parshall flume, based on the procedures given by Kilpatrick and Schneider (1983). The above equation will reproduce values of discharge precisely as shown in table 12, except for a few instances where the computed discharge deviates by 0.001 to 0.005 ft^3/s. This equation should not be used for values of gage height less than 0.01 ft or greater than 0.59 ft.

Table 12. Rating table for 3-inch modified Parshall flume.

[ft, feet; ft^3/s, cubic foot per second]

Gage height (ft)	Discharge (ft³/s)	Gage height (ft)	Discharge (ft³/s)	Gage height (ft)	Discharge (ft³/s)
0.01	0.0008	0.21	0.097	0.41	0.280
0.02	0.0024	0.22	0.104	0.42	0.290
0.03	0.0045	0.23	0.111	0.43	0.301
0.04	0.0070	0.24	0.119	0.44	0.312
0.05	0.010	0.25	0.127	0.45	0.323
0.06	0.013	0.26	0.135	0.46	0.334
0.07	0.017	0.27	0.144	0.47	0.345
0.08	0.021	0.28	0.153	0.48	0.357
0.09	0.025	0.29	0.162	0.49	0.368
0.10	0.030	0.30	0.170	0.50	0.380
0.11	0.035	0.31	0.179	0.51	0.392
0.12	0.040	0.32	0.188	0.52	0.404
0.13	0.045	0.33	0.198	0.53	0.417
0.14	0.051	0.34	0.208	0.54	0.430
0.15	0.057	0.35	0.218	0.55	0.443
0.16	0.063	0.36	0.228	0.56	0.456
0.17	0.069	0.37	0.238	0.57	0.470
0.18	0.076	0.38	0.248	0.58	0.483
0.19	0.083	0.39	0.259	0.59	0.497
0.20	0.090	0.40	0.269		

Volumetric Measurements

The most accurate way of measuring small discharges is the volumetric method. This method is performed by observing the time it takes to fill a container of known capacity, or the time required to partly fill a calibrated container to a known volume. The basic equipment needed for this method is a calibrated container and a stopwatch.

Two methods can be used to calibrate the container. The first method is to add known volumes of water by increments and note the depth of water in the container. The second method is to first weigh the empty container and then add varying amounts of water to it, each time weighing the container with water, and noting the depth of water in the container. The following equation can then be used to compute the volume of water corresponding to the depth that was read:

$$V = \frac{W_2 - W_1}{w}, \qquad (23)$$

where V volume of water in container, in cubic feet,
        W_2 weight of container with water, in pounds,
        W_1 weight of empty container, in pounds,
        w unit weight of water, 62.4 lb/ft^3.

Volumetric measurements of discharge are made with two types of conditions:

- When the flow is or can be concentrated so that all of it may be diverted into a calibrated container.

- When the depth of water flowing over broad-crested weirs and dams is small and volumetric-increment samples can be obtained.

Under the first condition, measurements are made at V-notch weirs at artificial controls where all the flow is in a notch or catenary, and at places where a small earth dam can be built and all the water diverted through a pipe of small diameter. Sometimes it is necessary to place a trough against the artificial control to carry the water from the control to the calibrated container. If a small dam is built, the stage behind the dam must be allowed to stabilize before the measurement is begun. The measurement is made three or four times to ensure error-free and consistent results.

Volumetric measurements are made under the second condition by catching a segment of the streamflow with a container having a known width of opening. Samples are taken at a number of locations across the dam or weir similar to procedures used for current-meter measurements. The flow rate of each sample is increased by the ratio of the subsection width to the sampled width to obtain a discharge rate for each subsection. The total discharge of the stream is the summation of the discharge rates of each subsection.

Float Measurements

Floats have limited use in streamgaging, but they can be used where the velocity is too low to obtain reliable measurements with the current meter, or where flood measurements are needed and the measuring structure has been destroyed or it is impossible to use a meter. Both surface floats and rod floats can be used. Surface floats may be almost anything that floats, such as wooden disks, partly filled bottles, oranges, or pumpkins. Floating debris or ice cakes may serve as natural floats. Rod floats are usually made of wood and weighted on one end so they will float upright in the stream. Rod floats are sometimes made in sections so their length can be adjusted to fit the stream depth; however, they should not touch the streambed.

Two cross sections are selected along a reach of straight channel for a float measurement. The cross sections should be far enough apart so that the time the float takes to pass from one cross section to the other can be measured accurately. A travel time of at least 20 seconds is recommended, but a shorter time can be used on small streams with high velocities, where it is impossible to select an adequate length of straight channel. The edge of water for both cross sections should be referenced to stakes (or other marker) on each bank. Those points will be used at a later date, when conditions permit, to survey cross sections of the measurement reach, and to obtain the distance between cross sections. The surveyed cross sections will be used to determine the average cross section for the reach.

Float measurements may sometimes be made through a reach extending from the upstream to the downstream side of a bridge. This kind of reach may be useful where velocity is very slow and velocity observations by current meter are not reliable.

The procedure for a float measurement is to distribute a number of floats uniformly over the stream width, noting the position of each with respect to the bank. They should be placed far enough upstream from the first cross section so they attain the velocity of the stream before they reach the first cross section. Use a stopwatch to time their travel between the two cross sections. As each float passes the second cross section, note its distance.

The velocity of the float is equal to the distance between the cross sections divided by the time of travel. The mean velocity of flow in the vertical is equal to the float velocity multiplied by a coefficient that is based on the shape of the vertical-velocity profile and relative depth of immersion of the float. A coefficient of about 0.85 to 0.88 is commonly used to convert surface velocity to mean velocity. The coefficient for rod floats varies from 0.85 to 1.00, depending upon the shape of the cross section, the length of the rod, and the velocity distribution.

The procedure for computing discharge is similar to that for a mechanical current-meter measurement. The discharge in each partial section is computed by multiplying the average area of the partial section by the mean velocity in the vertical for that partial section. The total discharge is equal to the sum of the discharges for all the partial sections.

Discharge measurements made with floats under favorable conditions may be accurate to within ±10 percent. Wind may adversely affect the accuracy of the computed discharge by its effect on the velocity of the floats, especially if velocity is very slow. If a poor reach is selected and not enough float runs are made, the results can be as much as 25 percent in error.

Indirect Discharge Measurements

During floods, it is frequently impossible or impractical to measure peak discharges when they occur. Roads may be impassable; structures from which current-meter measurements might have been made may be nonexistent, not suitably located, or destroyed; knowledge of the flood rise may not be available enough in advance to allow reaching the site near the time of the peak; the peak may be so sharp that a satisfactory current-meter measurement could not be made, even with a hydrographer present at the time; the flow of debris or ice can prevent use of a current meter; or personnel limitations might make it impossible to obtain direct measurements of high-stage discharge at numerous locations during a short flood period. Consequently, many peak discharges must be determined after the passage of the flood by indirect methods, such as slope-area, contracted-opening, flow-over-dam, or flow-through-culvert, rather than by direct current-meter measurement. Detailed descriptions of the procedures used in collecting field data and in computing the discharge are given by Benson and Dalrymple (1967), Dalrymple and Benson (1967), Bodhaine (1968), Matthai (1967), and Hulsing (1967), which are in book 3, chapters A1–A5, of the USGS Techniques and Methods series. Various computer programs are available for computing the discharge for indirect measurements.

Tracer Discharge Measurements

Measurement of discharge by this method depends on determination of the degree of dilution of an added tracer solution by the flowing water. A solution of a stable tracer, such as a fluorescent dye or a radioactive chemical, is injected into the stream at either a constant rate or all at once. The solution becomes diluted by the discharge of the stream. Measurement of the rate of injection, the concentration of the tracer in the injected solution, and the concentration of the tracer at a cross section downstream from the injection point permits the computation of stream discharge. The accuracy of the method critically depends upon complete mixing of the injected solution through the stream cross section before the sampling station is reached and upon no adsorption of the tracer on stream-bottom materials. The method is recommended only for those sites where conventional methods cannot be employed owing to shallow depths, extremely high velocities, or excessive turbulence. A detailed description of the procedures and equipment used in measuring discharge by a dye-dilution method is given by Kilpatrick and Cobb (1985).

References Cited

Benson, M.A., and Dalrymple, Tate, 1967, General field and office procedures for indirect measurements: U.S. Geological Survey Techniques of Water-Resources Investigations, book 3, chap. A1, 30 p. (Also available at *http://pubs.usgs.gov/twri/twri3-a1/.*)

Bodhaine, G.L., 1968, Measurement of peak discharge at culverts by indirect methods: U.S. Geological Survey Techniques of Water-Resources Investigations, book 3, chap. A3, 60 p. (Also available at *http://pubs.usgs.gov/twri/twri3-a3/.*)

Buchanan, T.J., and Somers, W.P., 1969, Discharge measurements at gaging stations: U.S. Geological Survey Techniques of Water-Resources Investigations, book 3, chap A8, 65 p. (Also available at *http://pubs.usgs.gov/twri/twri3a8/.*)

Dalrymple, Tate, and Benson, M.A., 1967, Measurement of peak discharge by the slope-area method: U.S. Geological Survey Techniques of Water-Resources Investigations, book 3, chap. A2, 12 p. (Also available at *http://pubs.usgs.gov/twri/twri3-a2/.*)

Fulford, J.M., and Sauer, V.B., 1986: Comparison of velocity interpolation methods for computing open-channel discharge, *in* Subitsky, S.Y. , ed., Selected papers in the hydrologic sciences: U.S. Geological Survey Water-Supply Paper 2290, p. 139–44. (Also available at *http://pubs.usgs.gov/wsp/wsp2290/.*)

Herschy, Reg, 1994, The analysis of uncertainty in the stage-discharge relation: Flow Measurement and instrumentation, v. 5, no. 3, p. 188–190.

Hulsing, Harry, 1967, Measurement of peak discharge at dams by indirect methods: U.S. Geological Survey Techniques of Water-Resources Investigations, book 3, chap. A5, 29 p. (Also available at *http://pubs.usgs.gov/twri/twri3-a5/.*)

International Organization for Standardization, 1997, Measurement of liquid flow in open channels—Velocity-area methods, ISO 748:1997E, 41 p.

International Organization for Standardization, 2003, Hydrometry—Echo sounders for water depth measurements: International Organization for Standardization ISO–CD_4366_ (E revised), 15 p.

Kennedy, E.J., 1984, Discharge ratings at gaging stations: U.S. Geological Survey Techniques of Water-Resources Investigations, book 3, chap. A10, 59 p. (Also available at *http://pubs.usgs.gov/twri/twri3-a10/.*)

Kilpatrick, F.A., and Cobb, E.D., 1985, Measurement of discharge using tracers: U.S. Geological Survey Techniques of Water-Resources Investigations, book 3, chap. A16, 52 p. (Also available at *http://pubs.usgs.gov/twri/twri3-a16/.*)

Kilpatrick, F.A., and Schneider, V.R., 1983, Use of flumes in measuring discharge: U.S. Geological Survey Techniques of Water-Resources Investigations, book 3, chap. A14, 46 p. (Also available at *http://pubs.usgs.gov/twri/twri3-a14/.*)

Lipscomb, S.W., 1995, Quality assurance plan for discharge measurements using broadband acoustic Doppler current profilers: U.S. Geological Survey Open-File Report 95–701, 7 p. (Also available at *http://pubs.usgs.gov/of/1995/ofr95-701/.*)

Matthai, H.F., 1967, Measurement of peak discharge at width contractions by indirect methods: U.S. Geological Survey Techniques of Water-Resources Investigations, book 3, chap. A4, 44p. (Also available at *http://pubs.usgs.gov/twri/twri3-a4/.*)

Morlock, Scott, E., 1996, Evaluation of acoustic Doppler current profiler measurements of river discharge: U.S. Geological Survey Water-Resources Investigations Report 95–4218, 37 p. (Also available at *http://pubs.usgs.gov/wri/wri95-4218/.*)

Mueller, D.S., and Wagner, C.R., 2009, Measuring discharge with acoustic Doppler current profilers from a moving boat: U.S. Geological Survey Techniques and Methods 3–A22, 72 p. (Also available at *http://pubs.water.usgs.gov/tm3a22.*)

Mueller, D.S., and Wagner, C.R., 2006, Application of the Loop method for correcting acoustic Doppler current profiler discharge measurements biased by sediment transport: U.S. Geological Survey Scientific Investigations Report 2006–5079, 18 p. (Also available at *http://pubs.usgs.gov/sir/2006/5079/.*)

Oberg, K.A., Morlock S.E., and Caldwell, W.S., 2005, Quality assurance plan for discharge measurements using acoustic Doppler current profilers: U.S. Geological Survey Scientific Investigations report 2005–5183, 35 p. (Also available at *http://pubs.usgs.gov/sir/2005/5183/.*)

Oberg, K.A., and Mueller, D.S., 2007, Validation of stream-flow measurements made with acoustic Doppler current profilers: Journal of Hydraulic Engineers, v. 133, no. 12, p. 1421–1432.

Pierce, C.H., 1941, Investigations of methods and equipment used in stream gaging, part 1, Performance of current meters in water of shallow depth: U.S. Geological Survey Water-Supply Paper 868–A, 35 p. (Also available at *http://pubs.usgs.gov/wsp/wsp868-a/.*)

Pierce, C.H., 1947, Equipment for river measurements—Structures for cableways: U.S. Geological Survey Circular 17, 38 p., 25 pl. (Also available at *http://pubs.usgs.gov/circ/circ17/.*)

R.D. Instruments, Inc., 1989, Acoustic Doppler current profilers, principles of operation—A practical primer: San Diego, Calif., R.D. Instruments, Inc., 52 p.

R.D. Instruments, Inc., 1995, Direct reading and self contained broadband acoustic Doppler current profiler technical manual for Firmware version 5.XX: San Diego, Calif., R.D. Instruments, Inc., 460 p.

R.D. Instruments, 1996, Principals of operation. A practical primer for broadband acoustic Doppler current profilers (2d ed.): San Diego, Calif., R.D. Instruments, Inc., 52 p.

Rantz, S.E., and others, 1982, Measurement and computation of streamflow: U.S. Geological Survey Water-Supply Paper 2175, v. 2, 631 p. (Also available at *http://pubs.usgs.gov/wsp/wsp2175/html/wsp2175_vol2.html.*)

Rehmel, M.S., Stewart, J.A., and Morlock, S.E., 2003, Tethered acoustic Doppler current profiler platforms for measuring streamflow: U.S. Geological Survey Open-File Report 03–237, 15 p. (Also available at *http://pubs.usgs.gov/of/2003/ofr03-237/.*)

Sauer, V.B., and Meyer, R.W., 1992, Determination of error in individual discharge measurements: U.S. Geological Survey Open-File Report 92–144, 21 p. (Also available at *http://pubs.usgs.gov/of/1992/ofr92-144/.*)

Simpson, M.R., 2002, Discharge measurements using a Broad-Band Acoustic Doppler Current Profiler: U.S. Geological Survey Open-File Report 01–01, 123 p. (Also available at *http://pubs.usgs.gov/of/2001/ofr0101/.*).

Simpson, M.R., and Oltmann, R.N., 1993, Discharge measurement system using an acoustic Doppler current profiler with applications to large rivers and estuaries: U.S. Geological Survey Water-Supply Paper 2395, 32 p. (Also available at *http://pubs.usgs.gov/wsp/wsp2395/.*)

Smoot, G.F., and Novak, C.E., 1968, Calibration and maintenance of vertical-axis type current meters: U.S. Geological Survey Techniques of Water-Resources Investigations, book 8, chap. B2, 23 p. (Also available at *http://pubs.usgs.gov/twri/twri8b2/.*)

Smoot, G.F., and Novak, C.E., 1969, Measurement of discharge by the moving-boat method: U.S. Geological Survey Techniques of Water-Resources Investigations, book 3, chap. A11, 22 p. (Also available at *http://pubs.usgs.gov/twri/twri3-a11/.*)

SonTek/YSI Corporation, 2002, SonTek/YSI FlowTracker Handheld ADV technical documentation: San Diego, Calif., SonTek/YSI Corporation, 2 p.

Wagner, C.R., 1995, Stream-gaging cableways: U.S. Geological Survey Techniques of Water-Resources Investigations, book 3, chap. A21, 56 p. (Also available at *http://pubs.usgs.gov/twri/twri3-a21/.*)

Young, K.B., 1950, A comparative study of mean-section and midsection methods for computation of discharge measurements: U.S. Geological Survey Open-File Report 53–277, 52 p.

www.ingramcontent.com/pod-product-compliance
Lightning Source LLC
Chambersburg PA
CBHW081509170526
45166CB00008B/2600